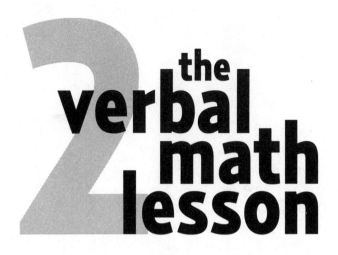

LEVEL 2
FOR CHILDREN AGES 4 TO 7

MICHAEL LEVIN, M.D.
CHARAN LANGTON. M S

Copyright 2008 Mountcastle Company
First Print Edition February 2008
First Electronic Edition June 2011
Second Edition April 2014

W0010583

Edited by Kelsey Negherbon, Ashley Kuhre and Julie Lundy
Design by Tijana Mihajlović

Manufactured in the United States of America

ISBN 978-0-913063-28-6

www.mathlesson.com

When a ball
Bounces off of a wall,
When you cook
From a recipe book,
When you know
How much money you owe,
That's mathematics!
When you choose
How much postage to use,
When you know
What's the chance it will snow,
When you bet
And you end up in debt,
Oh try as you may,
You just can't get away
From mathematics!
- That's Mathematics!

TOM LEHRER

INTRODUCTION

Verbal math, also called mental math, is a practical and time-honored method of solving mathematical problems. Math done with worksheets often slows children down. Shortcuts and computational tricks learned by doing math mentally allow children to bypass much of the tediousness they experience with written math.

The word problems in each lesson concentrate on one main concept so that the child can discover and apply the hidden mathematical pattern. The problems vary within each lesson to keep the child on his or her "mental toes" during the lesson. The language of the problems is kept at the level of a young child In this volume we teach addition and subtraction up to one-hundred and introduce multiplication and division. As in Verbal Math Lesson Book 1, no paper or pencil is needed. All the problems in this book are designed to be read to your child so they can be done verbally. This book is intended for the parent or teacher and not to be handed to the child. The answers are there because, well, adults are not as speedy as kids at math!

The Verbal Math Lesson should be done as a fun activity. You can do a few problems as time permits. Or you can create a daily program of about 10-15 minutes each day. It is not intended to replace school work or a more comprehensive math program.

We teach a different sequence of math operations than those used in school work. In Verbal Math we teach addition and subtraction along with multiplication and division simultaneously to demonstrate the reversibility of both practices. Objects or pictures are unnecessary for this course. However, from time to time you may need to illustrate a point by using objects or images. The decision to use these helpers is left to the parent. Many word problems in the Verbal Math Lesson books are accompanied by solutions. These solutions allow you to talk the child through new or harder problems and teach him or her how to think procedurally. The given solution should serve as a model for similar problems if your child asks for help.

There are twenty nine lessons in this book. Some children can do a lesson a week. For others, it might take longer. Sometimes repeating the lesson would be necessary before proceeding further. Just make sure that your child can do all the problems, and we mean all, correctly and speedily before going to the next lesson. You know your child better than anyone else. Trust your parental judgment and go as fast as your child allows. Although we suggest an age group for these books, these are just guidelines.

We hope you will contact us with your experience with this series and any suggestions for improvements and corrections.

Best of luck,
Michael Levin and Charan Langton

www.mathlesson.com

BOOK TWO

OPERATIONS WITH 2-DIGIT NUMBERS

PLACE VALUE

Let's learn place values:

Place value has to do with the place of a digit in a number. In number 62, for example, 6 is in the tens place. This is the second place counting from the right side of the number. It means that we have six 10s in this number. 2 is in the ones place which is always the right most place at the end of the number. This means that this number has 2 ones. If we write this out, the number 62 is the sum of 6, 10s and 2, 1s.

62 = 10 + 10 + 10 + 10 + 10 + 10 + 1 + 1

In a three-digit number, like 108, 1 is in hundreds place, zero is in tens, and 8 is in ones. So the number 108 is the sum of 1 100's and 8, 1s.

108 = 100 + 1 + 1 + 1 + 1 + 1 + 1 + 1 + 1

Tell me:

- In the number 29, in what place is 9? ***Ans:*** In the ones place.
- In the number 30, in what place is 0? ***Ans:*** In the ones place.

- In the number 81, in what place is 8? *Ans:* In the tens place.
- In the number 50, in what place is 5? *Ans:* In the tens place.
- In the number 109, in what place is 9? *Ans:* In the ones place.
- In the number 234, in what place is 3? *Ans:* In the tens place.
- In the number 234, in what place is 2? *Ans:* In the hundreds place.
- In the number 100, in what place is 1? *Ans:* In the hundreds place.
- In the number 341, in what place is 1? *Ans:* In the ones place.
- In the number 341, in what place is 4? *Ans:* In the tens place.

EXERCISE I

Addition Review: What is the sum or difference of the given numbers (in front of the equal sign) in these problems? Please do problems along columns.

33 + 11 = 44	23 + 27 = 50	35 + 45 = 80	22 + 30 = 52
42 + 15 = 57	19 + 11 = 30	42 + 48 = 90	31 + 29 = 60
45 + 15 = 60	21 + 32 = 53	33 + 66 = 99	30 + 50 = 80
46 + 24 = 70	22 + 33 = 55	34 + 36 = 70	
32 + 23 = 55	12 + 27 = 39	53 + 36 = 89	
34 + 26 = 60	25 + 35 = 60		

EXERCISE II

25 - 11 = 14	67 - 27 = 40	65 - 45 = 20	55 - 34 = 21
20 - 7 = 13	40 - 11 = 29	45 - 13 = 32	47 - 30 = 17
20 - 8 = 12	50 - 30 = 20	35 - 25 = 10	57 - 15 = 42
30 - 24 = 6	70 - 40 = 30	75 - 24 = 51	66 - 15 = 51
40 - 20 = 20	45 - 25 = 20	45 - 25 = 20	57 - 16 = 41
50 - 20 = 30	55 - 35 = 20	48 - 35 = 13	45 - 21 = 24

WORD PROBLEMS

1. A redwood tree is 80 feet tall and a fir tree is 60 feet tall. How much is the redwood tree taller than the fir tree? *Ans:* 20 feet.

2. A redwood tree is 80 feet tall and a birch tree is 30 feet tall. How much is the redwood tree taller than the birch tree? *Ans:* 50 feet.

3. A redwood tree is 40 feet tall and a fir tree is 30 feet tall. What is the height of the two trees together? *Ans:* 70 feet.

4. A redwood tree is 80 feet tall, a fir tree is 60 feet tall. By how much is the redwood tree taller than the fir tree? *Ans:* 20 feet.

5. Nancy paid $40 for groceries and has $60 left. How much money did she have before buying the groceries? *Ans:* $100.

6. Mina paid for $40 of groceries with a $50 bill. How much did she get in change? *Ans:* $10.

7. A carpenter opened a box which had 100 nails in it. He used 20 nails the first day and 30 the next day. How many nails are left in the box? *Ans:* 50 nails (100 - 20 = 80, then 80 - 30 = 50).

8. A plumber connected together three 20-foot pipes and then cut 10 feet from the end of the new pipe. How long is the new pipe? *Ans:* 50 feet (20 + 20 + 20 = 60; then 60 - 10 = 50).

9. An electrician spliced (joined together) a 30-foot wire and a 50-foot wire. How long is the new wire? *Ans:* 80 feet.

10. A roofer used 40 shingles on one side of the roof and 60 shingles on the other side. How many shingles did he use? *Ans:* 100 shingles.

11. A cable installer used 70 feet of cable from the street to the house and another 30 feet inside the house. How much cable did he use? *Ans:* 100 feet.

12. A butcher wrapped 20 ounces of beef and 70 ounces of lamb. How much meat did he wrap? *Ans:* 90 ounces.

13. A baker baked 50 cakes and 50 pies. How many of cakes and pies did she bake? *Ans:* 100 cakes and pies.

14. A cook pickled 40 cucumbers and 40 tomatoes. How many veggies did he pickle? *Ans:* 80 veggies.

15. Robert brought 2 checks to the bank, one for $30 and the other for $50. How much were both checks? *Ans:* $80.

16. A famous writer received 90 letters and wrote back 40. How many letters are waiting for his reply? *Ans:* 50 letters.

17. An absent-minded professor bought 40 pairs of glasses. He lost 20 pairs and broke 10. How many pairs of glasses are left for him break or lose? *Ans:* 10 pairs of glasses.
    Solution: 40 (pairs) - 20 (lost) = 20 pairs, then 20 (pairs) - 10 (broken) = 10 pairs left. The other way to solve this problem would be: 20 (lost) + 10 (broken) = 30 (lost and broken), then 40 (pairs) - 30 = 10 pairs.

18. There are 30 students on the tennis team and 40 students on the wrestling team.
    a) How many students are on both teams? *Ans:* 70 students.
    b) How many more students are on the wrestling team than on the tennis team? *Ans:* 10 more students.

19. The city hall building is 70 years old, the concert hall is 30 years old, and the school is 20 years old.
    a) By how much is the city hall older than the concert hall?
    Ans: 40 years.
    b) By how much is the city hall older than the school?
    Ans: 50 years.
    c) By how much is the concert hall older than the school?
    Ans: 10 years.

20. Mr. Handel can handle 30 candles. Mrs. Handel can handle 40 candles. How many candles can the two Handels handle?
    Ans: 70 candles.

21. There are 60 students, 20 teachers, and 10 custodians in a small school.
    a) How many people are in the school?
    Ans: 90 people (60 + 20 + 10 = 90).
    b) How many students and teachers are in the school?
    Ans: 80 students and teachers.
    c) How many more students are in the school than teachers?
    Ans: 40 more students.
    d) How many more students are in the school than teachers and custodians? *Ans:* 30 more students.

Solution: 20 (teachers) + 10 (custodians) = 30 people. Then, 60 (students) - 30 (teachers and custodians together) = 30 people.

22. Monday, Kirby read for 33 minutes. Tuesday, he read for 30 minutes. How many minutes did Kirby read on Monday and Tuesday? *Ans:* 63 minutes.

23. Lu shot 40 paint balls, Mia shot 44. How many paint balls did they both shoot? *Ans:* 84 paint balls.

24. Mia shot 21 paint balls out of the 41 she had. How many paint balls does she have left? *Ans:* 20 paint balls.

25. Before starting her training Ann could lift 40 pounds. Now she lifts 70 pounds. How many pounds more can Ann lift now? *Ans:* 30 pounds more.

26. Before training Ann weighed 80 pounds. Now she weighs 95 pounds. How many pounds did she gain? *Ans:* 15 pounds.

27. On a school trip to Florida, Pablo took 45 pictures of flamingos and 30 pictures of other kinds of birds. How many pictures of birds did he take? *Ans:* 75 pictures.

28. On the same trip, Mariana took 86 pictures. If she took 30 pictures of birds, how many other pictures did she take? *Ans:* 56 pictures.

29. An adult alligator has 80 teeth, a young alligator may have 40. How many more teeth does an adult alligator have? *Ans:* 40 teeth more.

30. The alligator in the zoo is 50 years old. How many more years might it live if a typical alligator can live up to 80 years? *Ans:* 30 years more.

31. For the 4th of July celebration we bought 85 red, white and blue balloons.
a) How many blue balloons did we buy if there are 50 red and white balloons? *Ans:* 35 blue balloons.
b) How many red balloons are there if white and blue balloons together make 60 balloons? *Ans:* 25 red balloons.

32. Carlos cut 53 stars and 40 circles from a gold poster board. How many figures did he cut out? ***Ans:*** 93 figures.

33. Carlos then cut 43 circles and 20 squares from a red poster board. How many figures did he cut now? ***Ans:*** 63 figures.

34. Mr. Wolff has a farm with 60 sheep and 28 pigs. How many animals are on Mr. Wolff's farm? ***Ans:*** 88 animals.

35. Mrs. Fox also has a farm with 47 rabbits and 50 chickens. How many animals are on Mrs. Fox's farm? ***Ans:*** 97 animals.

36. Sylvia and Lydia together have 98 quiz cards. How many cards does Sylvia have if Lydia has 60? ***Ans:*** 38 cards.

37. Two islands together have 83 trees. How many trees does the second island have if the first has 70 trees? ***Ans:*** 13 trees.

38. Ten years later, the second island that had 13 trees now has 53 trees. How many new trees grew on the island? ***Ans:*** 40 trees.

39. Two dogs together have 74 fleas. How many fleas does the second dog have, if the first dog has 30 flees? ***Ans:*** 44 fleas.

40. The second dog with the 44 fleas was washed by his owner and 30 fleas were washed off. How many fleas does he have left? ***Ans:*** 14 fleas.

41. The first puzzle has 30 pieces more than the second. How many pieces are in both puzzles if the second has 20 pieces? ***Ans:*** 70 pieces.
    Solution: The second puzzle has 20 pieces, the first puzzle has 20 + 30 = 50 (pieces). Both puzzles have 20 (in the first) + 50 (in the second) = 70 pieces.

42. Raj has $60 dollars. Sanjay has $40 dollars less. How much money do they have together? ***Ans:*** $80.
    Solution: Sanjay has $60 - $40 = $20. Both guys have $60 (Raj) + $20 (Sanjay) = $80 (together).

43. Fifty paintings were shown on the first day of a two-day art show, but 20 less paintings were shown the second day. How many paintings were shown the second day? ***Ans:*** 30 paintings.

44. An agent sold 40 tickets to a play, 30 tickets to a concert, and 20 tickets to a lecture. How many tickets did she sell? **Ans:** 90 tickets.

45. They are building a 70-story office building. So far they have built 30 floors. How many floors are left to build? **Ans:** 40 floors.

46. All of a sudden 100 yellow-jackets (wasps) appeared on Peter's patio; 40 of them came near enough to bite Peter. How many didn't come near? **Ans:** 60 yellow-jackets.

47. A baker made 30 cupcakes, 50 English muffins, and 60 scones.
a) How many more scones than muffins did he make?
Ans: 10 more.
b) How many cupcakes and muffins did he bake?
Ans: 80 altogether.
c) How many cupcakes and scones were baked?
Ans: 90 altogether.

48. Joe worked 30 hours the first week and 50 hours the second week.
a) How many more hours did Joe work the second week?
Ans: 20 more hours.
b) How many hours did Joe work in the two weeks?
Ans: 80 hours.

49. Anna has 11 dolls and Lydia has 3 more than Anna. How many dolls does Lydia have? **Ans:** 14 dolls.

50. Jackie's pre-school class has 20 children. Her Kindergarten class has 10 children more. How many children are in Jackie's kindergarten class? **Ans:** 30 children.

ADDING SINGLE-DIGIT NUMBERS TO DOUBLE-DIGIT NUMBERS

COUNTING

- Count aloud as fast as you can for 1 minute (by the clock). See how far you can count.

- Using a stopwatch (or looking at the second hand of a clock) count aloud as fast as you can by 10s (10, 20, 30, etc.). Check how long it took you and then ask you parent or teacher to match your time.

- Count from 1 to 50 skipping every other number (1, 3, 5, 7, 9, 11, etc.)

- Count from 0 to 50 skipping every other number (0, 2, 4, 6, 8, 10, etc.)

EXERCISE I

Let's compare double-digit numbers:

- Which number is larger, 54 or 45? *Ans:* 54

- Which number is smaller, 34 or 43? *Ans:* 34

- Which number is larger, 67 or 76? *Ans:* 76
- Which number is larger, 65 or 56? *Ans:* 65
- Which number is smaller, 78 or 87? *Ans:* 78
- Which number is larger, 89 or 98? *Ans:* 98
- Which number is smaller, 23 or 32? *Ans:* 23

Let's count:

- Count from 52 to 64 by adding 3 (i.e., 52, 55, 58, 61, 64).
- Count from 44 to 64 by adding 3 (i.e., 44, 47, 50, 53, 56, 59, 62, 65).
- Count from 71 to 91 by adding 4 (i.e., 71, 75, 79, 83, 87, 91).
- Count from 63 to 83 by adding 4 (i.e., 63, 67, 71, 75, 79, 83).
- Count from 41 to 66 by adding 5 (i.e., 41, 46, 51, 56, 61, 66).
- Count from 43 to 73 by adding 5 (i.e., 43, 48, 53, 58, 63, 68, 73).
- Count from 44 to 69 by adding 5 (i.e., 44, 49, 54, 59, 64, 69).
- Count from 30 to 60 by adding 6 (i.e., 30, 36, 42, 48, 54, 60).

EXERCISE II

22 + 6 = 28	42 - 7 = 35	38 + 8 = 46	70 - 24 = 46
37 - 17 = 20	44 + 6 = 50	56 - 9 = 47	45 + 25 = 70
30 + 6 = 36	45 - 7 = 38	21 + 8 = 29	
32 - 7 = 25	32 + 6 = 38	22 - 9 = 13	
33 + 3 = 36	30 - 7 = 23	34 + 7 = 41	

HOW TO SOLVE

First Problem: 48 + 5 = ?

Solution: Step 1: For number 48, it would take 2 to make it 50. We split 5 into 2 + 3.

Step 2: 48 + 2 = 50; then, 50 + 3 = 53.

The answer: 48 + 5 = 53.

Second Problem: 83 + 7 = ?

Solution: Seven does not need to be split to make to the nearest 10.

Third Problem: 73 + 8 = ?

Solution: Step 1:   8 splits into 7 + 1.

Step 2:   73 + 7 = 80; then, 80 + 1 = 81.

The answer: 73 + 8 = 81.

Fourth Problem: 89 + 5 = ?

Solution: Step 1:   5 splits into 1 + 4.

Step 2:   89 + 1 = 90, them 90 + 4 = 94.

The answer: 89 + 5 = 94.

TRICKS

Sometimes we need to learn tricks to help us solve math problems. Let's start with the simple trick of adding 9 to a number or adding a double-digit number to 9. We know that it is easier to add 10 and then take 1 away from a number. So we will use this idea.

45 + 9 = ?
We can pretend that we add 10 to 45 instead of 9, 45 + 10 = 55. But we added 1 to 9, now we have to take it back: 55 - 1 = 54. Then: 45 + 9 = 54.

9 + 73 = ?
We can use the same trick of turning 9 into 10 by adding 1, 10 + 73 = 83. Let's not forget to take back 1, 83 - 1 = 82. Then: 9 + 73 = 82.

EXERCISE III

22 + 6 = 28	42 + 6 = 48	38 + 7 = 45	70 + 17 = 87
37 + 7 = 44	44 + 8 = 52	56 + 8 = 64	45 + 18 = 63
30 + 8 = 38	45 + 7 = 52	21 + 7 = 28	48 + 19 = 67
32 + 9 = 41	32 + 6 = 38	22 + 6 = 28	52 + 16 = 68
33 + 7 = 40	30 + 8 = 38	34 + 16 = 50	47 + 17 = 64

WORD PROBLEMS

1.   An explorer found 57 islands, and then he found 9 more. How many islands did he discover? *Ans:* 66 islands.

2. My grandpa is 58 years old.
   a) How old is he going to be in 4 years? *Ans:* 62 years old.
   b) How old is he going to be in 5 years? *Ans:* 63 years old.
   c) How old is he going to be in 7 years? *Ans:* 65 years old.
   d) How old is he going to be in 9 years? *Ans:* 67 years old.

3. My grandma is 52 years old.
   a) How old was she 4 years ago? *Ans:* 48 years old.
   b) How old was she 5 years ago? *Ans:* 47 years old.
   c) How old was she 7 years ago? *Ans:* 45 years old.
   d) How old was she 9 years ago? *Ans:* 43 years old.

4. On a construction site, 8 workers were joined by 46 new ones. How many workers are on the site now? *Ans:* 54 workers.

5. The construction site had 56 workers, then 6 more were hired. How many workers are now at the site? *Ans:* 62 workers.

6. It is 67 miles to the last town and 8 more miles to the village. How many miles are there between the town and the village? *Ans:* 75 miles.

7. There were 56 cars on the parking lot, then 9 more cars came and parked. How many cars are there now in the parking lot? *Ans:* 65 cars.

8. A collector had 88 pins and bought 8 more. How many pins does she have now? *Ans:* 96 pins.

9. This summer, we bought 24 movie tickets and 9 concert tickets. How many tickets did we buy? *Ans:* 33 tickets.

10. The movie tickets for my mom and dad cost $18. My ticket cost $6. How much did all three tickets cost? *Ans:* $24.

11. We had three movie tickets that cost $24. Then my grandmother decided to come to the movie with us. If her ticket cost $7, how much money did we spend on all 4 tickets? *Ans:* $31.

12. The first game took 46 minutes; the second was 7 minutes longer. How long did the second game last? *Ans:* 53 minutes.

13. Think about a double-digit number less than 50. Add 3 to your number. Then add 6. Take away 9. You are back to your number!

14. A painter has 3 basic colors but can mix 88 more from the 3 basic colors. How many different colors can he have?
Ans: 92 colors.
The basic colors are red, blue, and yellow. You can make all other colors by mixing these. For example: green color comes from mixing blue and yellow, orange color comes from mixing red and yellow, purple comes from mixing red and blue, and so on.

15. There were 58 patients in a hospital,
a) After 5 new patients were admitted today, how many patients are in the hospital? *Ans:* 63 patients.
b) Then 9 more were admitted. How many patients are in the hospital now? *Ans:* 72 patients.
c) Then 8 more patients came. How many are there now? *Ans:* 80 patients.

16. Two pumps filled a gas tank. If the first pump pumped 8 gallons and the second pumped 25 gallons, how many gallons did both pumps pump? *Ans:* 33 gallons.

17. A far away sun has 7 planets and 17 asteroids. How many planets and asteroids are flying around that sun? *Ans:* 24 planets and asteroids.

18. A train was first delayed 45 minutes and then again for 8 more minutes. How long was the train delayed? *Ans:* 53 minutes.

19. Morris had 86 coins in his collection and then he purchased 9 more. How many coins are in his collection now? *Ans:* 95 coins. Did you know that the person who collects coins is called a numismatist?

20. If the sum of two numbers is 80 and one of the numbers is 8, what is the other number? *Ans:* 72.

21. Laura and Maya were jumping rope. Laura jumped 65 times and Maya jumped 8 times. How many times did they both jump? *Ans:* 73 times.

22. Peter and Ross are throwing snowballs. Peter threw 7 snowballs and Ross threw 88. How many snowballs did they both throw? *Ans:* 95 snowballs.

23. During the election, 45 candidates visited our town in September, but only 6 candidates in October. How candidates visited our town in those two months? *Ans:* 51 candidates.

24. It costs $47 to replace a broken door and $9 for a new lock. How much is the total cost of replacing the door? *Ans:* $56 dollars.

25. There are 26 letter keys and 10 number keys on the keyboard. How many letter and number keys altogether are on the keyboard? *Ans:* 36 keys.

26. There were 52 magazines in the library and they subscribed to 9 more magazines. How many magazines does the library get now? *Ans:* 61 magazines.
To subscribe to a magazine means to buy it so you get it regularly every month in the mail from the publisher.

27. In previous vacations we traveled through 24 states. This summer we plan to see 6 more. How many states will we have seen by the end of this summer? *Ans:* 30 states.

28. There were 33 explorers in the group before 7 new scouts joined in. How many people are in the group now?
Ans: 40 people.

29. It takes Lloyd 26 minutes to clean his room. It takes his brother 5 minutes longer to clean his. How long does it take Lloyd's brother to clean his room? *Ans:* 31 minutes.

30. One story is 57 pages long. The other story is 9 pages longer. How many pages are in the other story? *Ans:* 66 pages.

31. It takes 74 days to build a barn and 7 more days to paint it. How long does it take for the whole work? *Ans:* 81 days.

32. A man planned to drive 57 miles to his friend's house but got lost and had to drive around for 6 extra miles. How many miles did he travel? *Ans:* 63 miles.

33. A printer weighs 68 pounds and the paper inside weighs 5 lbs. How much do the printer and the paper weigh together?
Ans: 73 pounds.

34. Mr. Walker paid $28 for a dinner and left a $5 tip. How much did he pay altogether? *Ans:* $33.

35. Sally sells seashells by the seashore. She had 54 seashells and sold 8 of them. How many more seashells does she have to sell? *Ans:* 46 seashells.

36. An aspiring actress took 37 dancing lessons and 7 singing lessons. How many lessons did she take? *Ans:* 44 lessons. Aspiring means hoping and preparing for something desired.

37. Not counting the 6 pigeons, there are 47 birds in the park. How many birds are there if you count the pigeons? *Ans:* 53 birds.

38. It was 65 degrees at night and in the morning the temperature went up 7 degrees. What was the temperature in the morning? *Ans:* 72 degrees.

39. A truck drove 25 miles per hour on a town road, and 9 mph faster on a country road. What was the truck's speed on the country road? *Ans:* 34 mph.

40. A gardener counted 37 gopher holes and then noticed 9 more. How many gopher holes did the gardener see?
Ans: 46 gophers. Is the gardener happy?

41. Teresa had $77 in the bank and added $6 more. How much money does she have now? *Ans:* $83.

42. A professor gave 38 lectures in her country and 8 lectures abroad. How many lectures did she give? *Ans:* 46 lectures. Abroad means a foreign country.

43. After a group dance 8 main actors left the stage, but 38 dancers stayed on. How many actors and dancers were there at the start? *Ans:* 46.

44. Carlos could lift 44 pounds last year. This year he can lift 7 pounds more. How many pounds can he lift now? *Ans:* 51 lbs.

45. It took Nan 27 minutes to download 14 of her songs and 8 more minutes to download 5 more songs. How long did it take her to download all the songs? *Ans:* 35 minutes.
Solution: The reason this problem might be a challenge is because it has numbers that have nothing to do with the

answer. The question does not ask for the number of songs, though this information is in the problem, but asks only for the amount of time it took to download them. We can make it simpler by paying attention only to the time spent, that is, it took 27 minutes and then, 8 more minutes. Then the problem becomes 27 + 8 = 35 minutes.

46. I saw 25 apples in 5 baskets and 9 oranges in 3 baskets. How many baskets did I see? *Ans:* 8 baskets. Did you do it right? Bravo! if you did!

47. There are 74 pages in the first 6 chapters of the book and 8 pages in the 7ᵗʰ chapter. How many pages are in all seven chapters? *Ans:* 82 pages.
    Solution: 74 pages + 8 pages (in the seventh chapter) = 82 pages.

48. 16 zebras have 48 legs and 4 pelicans have 8 legs.
    a) How many zebras and pelicans are in the zoo?
    Ans: 20 zebras and pelicans.
    b) How many legs do all zebras and pelicans have?
    Ans: 56 legs.

49. Lee counted 44 sides on 11 squares and 9 sides on 3 triangles.
    a) How many sides did Lee count? *Ans:* 53 sides.
    b) How many squares and triangles were there?
    Ans: 14 squares and triangles.

50. Jessica bought 26 cupcakes for her friends. Then she remembered to buy cupcakes for her family and bought 9 more. How many did she buy in all? *Ans:* 35 cupcakes.

51. Last week Sevi gave 11 gold stars to Shawn because he was such a good boy. This week he got 7 more gold stars. Shawn needs 20 stars to get a prize. How many more gold stars does he need to get his prize? *Ans:* Only 2 more stars to go.

52. Shawn has 13 silver stars so far but needs 20. How many stars does he need to get a silver star prize? *Ans:* 7 more stars to go.

53. Bindi did 24 problems from the lesson so far. She still has 11 more problems left to do. How many problems are in the lesson? *Ans:* 35 lbs.

SUBTRACTING SINGLE-DIGIT NUMBERS FROM DOUBLE-DIGIT NUMBERS

SKIP COUNTING

- Count from 60 to 100, skipping every other number (i.e., 60, 62, 64, 66, 68, 70)
- Count from 51 to 99 skipping every other number (i.e., 51, 53, 55, 57, 59, 61)
- Count from 70 down to 40 skipping every other number (i.e., 70, 68, 66, 64, 62, 60).
- Count from 81 down to 59 skipping every other number (i.e., 81, 79, 77, 75, 73, 71).

EXERCISE I

56 - 6 = 50	82 - 7 = 75	71 - 9 = 62	45 - 7 = 38
37 - 7 = 30	45 - 8 = 37	42 - 3 = 39	62 - 8 = 54
45 - 8 = 37	76 - 9 = 67	63 - 6 = 57	
67 - 9 = 58	36 - 4 = 32	64 - 9 = 55	
33 - 3 = 30	39 - 5 = 34	58 - 6 = 52	
57 - 6 = 51	41 - 8 = 33	91 - 5 = 86	

HOW TO SOLVE

First Problem: 34 - 5 = ?

An easy way to solve this problem is by splitting 5 into 4 and 1 (because 4 + 1 = 5) and then to subtract them one at a time.

Solution: Step 1: 5 = 4 + 1

Step 2: 34 - 4 = 30; next 30 - 1 = 29

The answer: 34 - 5 = 29.

Second Problem: 55 - 8 = ?

Solution: Step 1: 8 = 5 + 3

Step 2: 55 - 5 = 50; next 50 - 3 = 47

The answer: 55 - 8 = 47.

Third Problem: 74 - 6 = ?

Solution: Step 1: 6 = 4 + 2

Step 2: 74 - 4 = 70; next 70 - 2 = 68

The answer: 74 - 6 = 68.

Remember that when doing subtraction you can't switch numbers around and 74 - 6 is *not* equal to 6 - 74!

EXERCISE II

56 - 9 = 47	76 - 9 = 67	83 - 18 = 65	67 + 9 = 76
37 - 9 = 28	36 + 8 = 44	55 - 19 = 36	54 - 9 = 45
45 + 9 = 54	39 - 9 = 30	58 + 8 = 66	76 - 9 = 67
67 - 9 = 58	41 + 19 = 60	91 - 9 = 82	63 + 9 = 72
33 + 9 = 42	71 - 19 = 52	31 - 9 = 22	
57 - 9 = 48	42 + 19 = 61	45 + 9 = 54	
82 + 9 = 91	63 - 19 = 44	62 - 9 = 53	
45 - 8 = 37	64 + 18 = 82	82 - 9 = 73	

TRICKS

Let's learn a new trick for subtracting 9 from a number.

Problem: 45 - 9 = ?

We can pretend that we will take away 10 instead of 9. Then 45 - 10 = 35. But we took away 10, which is 1 more than 9, so we have to give 1 back: 35 + 1 = 36.

The answer: 45 - 9 = 36.

EXERCISE III

56 - 8 = 48	76 - 18 = 58	80 - 26 = 54	63 + 13 = 76
37 - 7 = 30	46 + 14 = 60	45 - 27 = 18	54 - 4 = 50
45 + 5 = 50	43 - 13 = 30	35 + 28 = 63	14 - 6 = 8
67 - 14 = 53	32 + 15 = 47	32 - 24 = 8	19 + 6 = 25
33 + 11 = 44	71 - 16 = 55	38 - 11 = 27	28 - 8 = 20
57 - 13 = 44	67 + 14 = 81	44 + 15 = 59	39 - 9 = 30
82 + 16 = 98	38 - 21 = 17	62 - 13 = 49	49 + 3 = 52
45 - 17 = 28	59 + 22 = 81	68 - 14 = 54	

WORD PROBLEMS

1. After Cathy paid $44 for an electric drill, she had $4 left in her wallet. How much money did she have before the purchase?
    Ans: $48 dollars.
    Solution: Before she paid for the drill, Cathy had $44 (the price of the drill) + $4 (left in the wallet) = $48.

2. At 8 o'clock, 55 taxicabs left the parking lot but 5 stayed behind. How many cabs were on the lot before 8 o'clock?
    Ans: 60 taxicabs.

3. If my uncle is 56 and my aunt is 5 years younger, how old is my aunt? *Ans:* 51 years old.

4. My aunt is 51 years old.
    a) How old was she 2 years ago? *Ans:* 49 years old.
    b) How old was she 4 years ago? *Ans:* 47 years old.
    c) How old was she 6 years ago? *Ans:* 45 years old.
    d) How old was she 8 years ago? *Ans:* 43 years old.

5. Vikram spelled 44 words correctly in a game. Tina spelled correctly 5 words fewer. How many words did Tina spell correctly? *Ans:* 39 words.

6. Bella used 31 roses in two bouquets. If one bouquet had 7 roses, how many did the second bouquet have? *Ans:* 24 roses.

7. If 43 flies flew in the room and 8 flies flew out, how many flies are still in the room? *Ans:* 35 flies.

8. If out of the 35 flies in the room, 8 were caught by a spider, how many flies are still flying around? *Ans:* 27 flies.

9. If out of 27 flies only 8 found the pie sitting on the table, how many went away hungry? *Ans:* 19 flies.

10. Petra divided 35 presents into two boxes, 8 presents in one box, the rest in the other. How many presents were in the second box? *Ans:* 27 presents.

11. Miriam cut 6 feet from a 32-foot rope. How much rope is left? *Ans:* 26 feet.

12. The chef cut a chicken into 34 pieces. He fried 9 and froze the rest. How many pieces did he freeze? *Ans:* 25 pieces.

13. It took 43 weeks for the preparation and the trip. The trip lasted 8 weeks. How long did it take to prepare? *Ans:* 35 weeks.

14. Jerry picked 54 tomatoes and threw away 8. How many did he keep? *Ans:* 46 tomatoes.

15. Preston opened a bag with 54 jelly beans and immediately ate 9. How many are left in the bag? *Ans:* 45 jelly beans.

16. A shirt and a tie cost $43. The tie is $8. How much is the shirt? *Ans:* $35.

17. 51 days are left before Marla's birthday.
    a) How many days will be left after 3 days? *Ans:* 48 days.
    b) How many days will be left after 5 days? *Ans:* 46 days.
    c) How many days will be left after 7 days? *Ans:* 44 days.
    d) How many days will be left after 9 days? *Ans:* 42 days.

18. A theater auditioned 62 actors and selected 5. How many got rejected? *Ans:* 57.

19. Leah picked 70 plums and ate 9 of them. How many did she take home? *Ans:* 61 plums.

20. Trent bought 60 postcards and mailed 7. How many did he keep? *Ans:* 53 postcards.

21. Freddy gave the cashier a $50 bill and got back $6 in change. How much did he spend? *Ans:* $44.

22. I cut off 4 feet from a 60 foot ribbon. How much is left? *Ans:* 56 feet.

23. Out of 53 students 7 failed the test. How many passed? *Ans:* 46 students.

24. Out of a 64-ounce bottle of juice, Jim poured out 8 ounces. How many ounces are left? *Ans:* 56 ounces.

25. Out of 71 light bulbs on the garland, 9 bulbs burned out. How many light bulbs are lit? *Ans:* 62 light bulbs.

26. Mr. Citrus received a box with 63 oranges and threw away 7 rotten ones. How many did he keep? *Ans:* 56 oranges.

27. Mrs. Fisher caught 56 trout but had to throw away 7. How many did she keep to sell? *Ans:* 49 trout.

28. Mr. Musak wrote 49 songs but didn't like 7. How many of his songs did he like? *Ans:* 42 songs.

29. Dr. Needles saw 42 patients and gave shots to 7 of them. How many were spared? *Ans:* 35 shots.

30. Erik Ball pitched a ball 54 times and struck out 6 times. How many times was the ball hit? *Ans:* 48 times.

31. Mr. Bishop played 48 chess games and won 6. How many did he lose or tied? *Ans:* 42 games.

32. Lady Target shot her bow 42 times and hit the bulls eye 6 times. How many times did she miss the target? *Ans:* 36 times.

33. Mr. Taylor mended 80 shirts and pants. There were 8 shirts. How many pants did he mend? *Ans:* 72 pants.

34. Rory weighed 66 pounds and gained 7 more. How much does she weigh now? *Ans:* 73 pounds.

35. Rory's sister Cory who was 93 pounds, got sick and lost 5 pounds. What's Cory's weight now? *Ans:* 88 pounds.

36. Betty wrote her name 88 times and then 8 times more. How many times did she write her name? *Ans:* 96 times.

37. A ship carried 52 people out of which 9 were crew members. How many passengers were on the ship? *Ans:* 43 passengers.

38. There were 90 monkeys in a forest. One day they saw a python and 9 monkeys decided to move to a far away forest. How many stayed? *Ans:* 81 monkeys.

39. Out of 60 seeds planted, only 7 germinated. How many didn't? *Ans:* 53 seeds. Germinated means to grow from seed.

40. In 50 years it snowed during only 8 winters in my town. How many years it didn't snow? *Ans:* 42 years.

41. In a battle with a monster jelly fish, a shark lost 17 out of her 100 teeth. How many teeth does she have left? *Ans:* 83 teeth.

42. Jay broke a big chocolate bar into 40 pieces and gobbled up 8 pieces right away. How many pieces did he leave for later? *Ans:* 32 pieces of chocolate.

43. One day in October there were 52 acorns on an oak tree. The wind blew away 8. How many acorns are still hanging on the tree? *Ans:* 44 acorns.

44. Fran counted 53 batteries in the box. If there were 8 large batteries, how many small ones were in the box? *Ans:* 45 batteries.

45. When my dad opened his presents, there were 8 ties. He put them in his closet together with the 55 ties he received before. How many ties does he have now? *Ans:* 63 ties.

46. Zephyr, a small cruise ship, has 42 people aboard. If 9 are crew members, how many passengers are on the Zephyr? *Ans:* 33 passengers.

47. Chloe spent $53 on a new guitar. Of this amount $5 was tax. What was the price of the guitar? *Ans:* $48.

48. There are 36 flowers in a vase which can hold 7 more. How many flowers can fit in this vase? *Ans:* 43 flowers.

49. Become-a-Famous-Clown course lasted 61 days. There were 8 weekend days, and the rest were class days. How many class days were there? **Ans:** 53 days.

50. There were 73 beans in a jar. After Liam dropped the jar, there were only 9 beans left in it. How many beans spilled on the floor? **Ans:** 64 beans.

51. A television show lasted 60 minutes, 9 minutes of which were for commercials. How much time was left for the show? **Ans:** 51 minutes.

52. A nursery sold 44 trees, 9 of which were birch trees; the rest were pines. How many pines did they sell? **Ans:** 35 pines.

53. Murray had 36 coins in his collection. If he gave away 8, how many coins are left? **Ans:** 28 coins.

54. A truck had 39 packages after delivering 7. How many packages did the driver start with? **Ans:** 46 packages.

55. Staying home from school with flu, Omar sent 34 text messages to his classmates and received 6 replies. How many messages went unanswered? **Ans:** 28 messages.

56. Alan had 36 coins in his collection. If he gave away 5, how many coins are left? **Ans:** 31 coins.

57. Alan then had 31 coins in his collection. If he gave away 8, how many coins are left? **Ans:** 23 coins.

58. Alan then had 23 coins in his collection. If he gave away 7, how many coins are left? **Ans:** 16 coins.

59. Alan then had 16 coins in his collection. If he gave away 8, how many coins are left? **Ans:** 8 coins.

60. Alan's mom asked him, "How many coins did you give away so far?" Alan said, "First I gave away 5 coins, then 8, then 7, and then 8 again." How many coins did Alan give away? **Ans:** 28 coins.

61. Alan's mom said to him, "You had 36 coins and now you only have 8 coins." What's the other way of figuring out how many coins Alan gave away? **Ans:** Subtract 8 from 36, 28 coins.

ADDING DOUBLE-DIGIT NUMBERS ENDING IN 1 OR 2

SKIP COUNTING

- Count from 60 to 87 skipping 2 numbers or, in other words, adding 3 (i.e., 60, 63, 66, 69, 72, 75, etc.)

- Count from 61 to 88 skipping 2 numbers or, in other words, adding 3 (i.e., 61, 64, 67, 70, 73, 76, etc.)

- Count from 70 down to 40 skipping 2 numbers or, in other words, subtracting 3 (i.e., 70, 67, 64, 61, 58, 55, etc.).

- Count from 72 down to 59 skipping 2 numbers or, in other words, subtracting 3 (i.e., 72, 69, 66, 63, 60, 57, etc.).

ADDING DOUBLE-DIGIT NUMBERS

Adding double-digit numbers can be easier if we first split the second number into tens and ones. Next, we add the tens from the second number to the first number. Then (as a second step) we add the ones.

Please remember that the exact order does not matter; the second number can switch places with the first and vice versa.

EXERCISE I

20 + 10 = 30	30 + 60 = 90	100 - 20 = 80	20 + 50 = 70
40 + 30 = 70	90 - 50 = 40	100 - 40 = 60	90 - 60 = 30
40 - 30 = 10	70 - 50 = 20	70 - 70 = 0	40 + 40 = 80
50 + 30 = 80	20 + 60 = 80	50 + 40 = 90	30 + 70 = 100
70 + 20 = 90	10 + 90 = 100	50 - 40 = 10	50 - 10 = 40
70 - 20 = 50	30 + 30 = 60	20 + 80 = 100	30 + 40 = 70

EXERCISE II

40 + 20 = 60	33 + 20 = 53	33 + 50 = 83	15 + 70 = 85
12 + 10 = 22	40 + 22 = 62	20 + 49 = 69	40 + 48 = 88
12 + 20 = 32	71 + 10 = 81	10 + 89 = 99	36 + 20 = 56
18 + 20 = 38	55 + 30 = 85	71 + 20 = 91	84 + 10 = 94
25 + 20 + 45	67 + 30 = 97	35 + 40 = 75	28 + 50 = 78

EXERCISE III

22 + 30 = 52	32 + 23 = 55	35 + 43 = 78	19 + 20 = 39
31 - 20 = 11	30 + 26 = 56	77 - 44 = 33	19 + 31 = 50
30 + 50 = 80	38 - 27 = 11	33 + 66 = 99	67 - 27 = 40
32 + 11 = 43	56 - 11 = 45	34 + 36 = 70	15 + 45 = 60
33 - 11 = 22	21 + 32 = 53	56 - 43 = 13	58 - 37 = 21
42 + 15 = 57	22 + 33 = 55	11 + 56 = 67	68 - 42 = 26
44 + 15 = 59	44 - 27 = 17	13 + 44 = 57	29 + 51 = 80
45 - 24 = 21	25 + 35 = 60	87 - 41 = 46	44 + 22 = 66

HOW TO SOLVE

First Problem: 35 + 11 = ?

Solution: Step 1: Let's break the second number into 11 = 10 + 1.

Step 2: Next, 35 + 10 = 45, and then 45 + 1 = 46.

The answer: 35 + 11 = 46.

Second Problem: 27 + 22 = ?

Solution: Step 1: Let's break the second number into 22 = 20 + 2.

Step 2: Next, 27 + 20 = 47, and then 47 + 2 = 49.

The answer: 27 + 22 = 49.

Third Problem: 49 + 41 = ?

Solution: Step 1:   41 = 40 + 1.

Step 2:   Next, 49 + 40 = 89, and then 89 + 1 = 90.

The answer: 49 + 41 = 90.

Fourth Problem: 59 + 22 = ?

Solution: Step 1:   22 = 20 + 2.

Step 2:   Next, 59 + 20 = 79, and then 79 + 2 = 81.

The answer: 59 + 22 = 81.

WORD PROBLEMS

1.   Mr. Penny wrote 60 checks and then had to write 21 more checks. How many checks did he write? *Ans:* 81 checks.

2.   To help the 40 sentries (guards) on duty, the commander assigned 22 more. How many sentries are on duty? *Ans:* 62 sentries.

3.   May was sweeping the floor for 34 minutes and cleaning the kitchen for 12 minutes. How many minutes did May work? *Ans:* 46 minutes.

4.   If there were 44 sail boats in the bay and 32 more boats sailed in, how many boats are in the bay now? *Ans:* 76 sail boats.

5.   A genius learned the Chinese language in 37 days and Arabic in 32 days. How long did the genius take to learn both languages? *Ans:* 69 days.

6.   Mr. Macroni let the phone ring 22 times the first time and 28 times before he answered the second time. How many times did the phone ring? *Ans:* 50 times.

7.   Michael's shirt cost $22. His hat cost $18 more than the shirt. How much did the hat cost? *Ans:* $40.

8.   A fish weighs 38 pounds but a lobster is only 12 pounds. How much do they both weigh together? *Ans:* 50 pounds.

9. Irene drew a picture with 35 trees and 21 bushes. How many trees and bushes did she draw? **Ans:** 56 trees and bushes.

10. A chef wrote a cook book with 43 recipes for pasta dishes and 32 for rice. How many recipes are in the book? **Ans:** 75 recipes.

11. A plumber cut a pipe into two pieces; one was 24 inches long and the other was 11 inches long. How long was the pipe before the plumber cut it? **Ans:** 35 inches.

12. A farmer plowed 35 acres and his daughter plowed 31. How many acres did they both plow? **Ans:** 66 acres.

13. There are 32 miles to the beach. How long is the round trip? **Ans:** 64 miles (32 + 32 = 64).

14. One school took 47 children and the other took 21. How many children were taken by both schools? **Ans:** 68 children.

15. Two friends walked in the opposite directions. One walked 22 miles and the other walked 21 miles. How far apart are they? **Ans:** 43 miles.

16. An annoying fly flew by 31 times from left to right and 32 times from right to left. How many times did the fly fly by? **Ans:** 63 times.

17. After hugging a cactus, Nia took out 26 needles from her right arm and 32 needles from the left. How many needles did the cactus give her? **Ans:** 58 needles.

18. While investigating a fruit theft, the detective found 49 cherry pits and 21 plum pits behind the building. How many fruits did the thieves eat? **Ans:** 70 fruits.

19. At first there were 37 Merry Men in Robin Hood's band. Then 22 more joined. How many Merry Men are in the band now? **Ans:** 59 Merry Men.

20. Roy put 33 logs in the rack, and then added 11 more. How many logs did Roy put in the rack? **Ans:** 44 logs.

21. One alien has 24 ears and the other has 32 more ears. How many ears does the second alien have? **Ans:** 56 ears.

22. The road goes 62 miles up the hill and 17 miles down the hill. How long is the road? *Ans:* 79 miles.

23. After telling a lie, Pinocchio saw his nose grow 23 inches. After telling another lie, his nose grew 11 inches more. How much did his nose grow altogether? *Ans:* 34 inches.

24. A mountain goat jumped 15 feet across a stream and then 20 feet across some rocks. How long were the two jumps together? *Ans:* 35 feet.

25. A farmer caught 31 mice and 18 squirrels. How many animals did the farmer catch? *Ans:* 49 animals.

26. There are 36 miles to the next rest stop, and the gas station is 12 miles further. How far are we from the gas station? *Ans:* 48 miles.

27. A porter carried 54 suitcases and 31 duffel bags upstairs. How many pieces of luggage did he carry altogether? *Ans:* 85 pieces.

28. Inside the pool, the water temperature is 68 degrees. Outside, it is 12 degrees higher. What is the outside temperature? *Ans:* 80 degrees.

29. If the baby took 38 steps and then took 21 more steps, how many steps did the baby take? *Ans:* 59 steps.

30. If a tree lost 35 leaves,
    a) How many leaves would it have lost after 11 more leaves fell down? *Ans:* 46 leaves.
    b) How many leaves would it have lost after 21 more leaves fell down? *Ans:* 56 leaves.
    c) How many leaves would it have lost after 31 more leaves fell down? *Ans:* 66 leaves.
    d) How many leaves would it have lost after 12 more leaves fell down? *Ans:* 47 leaves.

31. First, 55 rats followed the Pied Piper of Hamelin, and then 32 more rats joined. How many rats followed the Piper altogether? *Ans:* 87 rats. Do you know the story of the Pied Piper of Hamelin?

32. The price of oil was $37 per barrel. Then it went up $12 per barrel. What is the oil price now? *Ans:* $49 per barrel.

33. A computer engineer used 23 transistors and 32 diodes to makes the circuit board. How many parts did he use? *Ans:* 55.

34. A girl is 23 years old, which is 12 years younger than her brother. How old is her brother? *Ans:* 35 years old.

35. In a shop, there were 27 bottles of shampoo and 31 bottles of hair conditioner. How many bottles of both were in the shop? *Ans:* 58 bottles.

36. At the party, 22 guests wore bow ties and 27 guests wore regular ties.
    a) How many people wore ties? *Ans:* 49 people.
    b) If 31 guests wore no ties, how many guests were with and without ties altogether? *Ans:* 80 people.

37. At the party 32 guests ordered chicken and 23 ordered lamb.
    a) How many people ordered meat? *Ans:* 55 people.
    b) If 21 people ordered fish, how many had either chicken or fish? *Ans:* 53 people.

38. After buying 42 new combs, Rapunzel remembered that she already had combs. How many combs does Rapunzel have now? *Ans:* 78 combs.

39. A fisherman sold two fish. One for $44 and the other for $42. How much money did he get? *Ans:* $86.

40. A 19-foot blue line is attached to a 2-foot long red line. How long are both lines together? *Ans:* 21 feet.

41. During the hockey game, a goalie held off 29 shots and missed 2. How many shots were made? *Ans:* 31 shots.

42. A city divided the park into two parts, 29 acres for one and 12 acres for the other. How big is the park? *Ans:* 41 acres.

43. If Peter, who weighs 42 pounds, can hold his 9 pound pet python around his shoulders, how much will they weigh together? *Ans:* 51 pounds.

44. If Peter, who weighs 42 pounds, holds his 19 pound pet skunk in his arms, how much do they weigh together?
Ans: 61 pounds.

45. If Leslie sprayed mom's perfume on 17 girls and 22 boys before she got in trouble, how many kids were smelling really nice?
Ans: 39 kids.

46. Tracy hit a golf ball 39 feet and then 12 more feet to the hole. How many feet did the ball go? *Ans:* 51 feet.

47. One fish the fisherman caught weighed 11 pounds the other fish weighed 19 pounds. How much did they both weigh?
Ans: 30 pounds.

48. To a 29-car train they added 22 more cars. How many cars are there now? *Ans:* 51 cars.

49. Before Keith added $32 to his savings he had $67 in his account. How much money is in his savings account now?
Ans: $99.

50. Missy has 42 freckles. Her twin brother Nick has 39 freckles. How many freckles do the twins have? *Ans:* 81 freckles.

51. Tony, the dog, chased 28 finches and 12 pigeons. How many birds did he chase? *Ans:* 40 birds.

SUBTRACTING DOUBLE-DIGIT NUMBERS ENDING IN 1 OR 2

SKIP COUNTING

- Count from 50 down to 0 skipping every other number (i.e., 50, 48, 46, 44, etc.)

- Count from 51 to 1 skipping every other number (i.e., 51, 49, 47, 45, 43, etc.)

- Count from 90 down to 69 by skipping 2 numbers or, in other words, subtracting 3 (i.e., 90, 87, 84, 81, 78, 75, 72, 69).

- Count from 62 down to 38 by skipping 3 numbers or, in other words, subtracting 4 (i.e., 62, 58, 54, 50, 46, 42, 38).

EXERCISE I

21 - 11 = 10	33 - 12 = 21	65 - 22 = 43	77 - 22 = 55
31 - 11 = 20	44 - 12 = 32	56 - 22 = 34	58 - 41 = 17
42 - 11 = 31	54 - 21 = 33	66 - 41 = 25	88 - 32 = 56
42 - 21 = 21	46 - 32 = 14	67 - 32 = 35	94 - 62 = 32
33 - 22 = 11	56 - 31 = 25	76 - 22 = 54	98 - 61 = 37

49 - 21 = 28	36 - 9 = 27	63 - 8 = 55	71 - 12 = 59
25 - 8 = 17	15 - 8 = 7	69 - 52 = 17	73 - 14 = 59
55 - 8 = 47	45 - 8 = 37		

HOW TO SOLVE

First Problem: 43 - 11 = ?

Solution: Step 1: Let's break number 11 into 10 and 1.
Step 2: Then, 43 - 10 = 33, and 33 - 1 = 32.

The answer: 43 - 11 = 32.

Second Problem: 50 - 21 = ?

Solution: Step 1: We'll break 21 into 20 and 1.
Step 2: Then, 50 - 20 = 30, next 30 - 1 = 29.

The answer: 50 - 21 = 29.

Third Problem: 61 - 32 = ?

Solution: Step 1: 32 = 30 + 2
Step 2: Then, 61 - 30 = 31, next 31 - 2 = 29.

The answer: 61 - 32 = 29.

Fourth Problem: 90 - 62 = ?

Solution: Step 1: 62 = 60 + 2
Step 2: Then, 90 - 60 = 30, next 30 - 2 = 28

The answer: 90 - 62 = 28.

EXERCISE II

21 - 12 = 9	51 - 42 = 9	81 - 62 = 19	19 - 10 = 9
31 - 12 = 19	60 - 21 = 39	50 - 12 = 38	19 - 9 = 10
40 - 11 = 29	51 - 32 = 19	90 - 32 = 58	
61 - 22 = 39	81 - 52 = 29	80 - 21 = 59	
80 - 12 = 68	91 - 62 = 29	19 - 2 = 17	
80 - 42 = 38	70 - 32 = 38	19 - 11 = 8	

WORD PROBLEMS

1. Thirty-four stray cats gathered on a roof for a sing-along. After a while, 11 cats left because they were bored. How many cats stayed? *Ans:* 23 cats.

2. A sunflower had 58 seeds.
   a) After 11 seeds fell out, how many seeds were left in the flower? *Ans:* 47 seeds.
   b) After 21 seeds fell out, how many seeds were left in the flower? *Ans:* 37 seeds.
   c) After 31 seeds fell out, how many seeds were left in the flower? *Ans:* 27 seeds.
   d) After 41 seeds fell out, how many seeds were left in the flower? *Ans:* 17 seeds.
   e) After 51 seeds fell out, how many seeds were left in the flower? *Ans:* 7 seeds.

3. Out of 63 students in a small school, 21 went on a field trip. How many stayed in school that day? *Ans:* 42 students.

4. A flower shop sold 56 roses and 11 peonies. How many flowers did the shop sell? *Ans:* 67 flowers.

5. Out of a 42-minute exercise class, we ran for 22 minutes and rested the rest of the time. How long did we rest? *Ans:* 20 minutes.

6. There are 52 cards in the deck.
   a) If I take 12 cards out, how many cards remain? *Ans:* 40 cards.
   b) If I take 22 cards out, how many cards remain? *Ans:* 30 cards.
   c) If I take 32 cards out, how many cards remain? *Ans:* 20 cards.

7. Out of 70 days of summer vacation, Meena tutored for 22 days. How many days she did she not tutor? *Ans:* 48 days.

8. Before the party, Mrs. Dulcet counted 41 cubes of sugar in the sugar bowl. After the party, there were only 11 cubes. How many lumps of sugar got used? *Ans:* 30 cubes.

9. Before the party, Mrs. Dulcet counted 41 muffins on a tray. Afterwards, only 30 remained. How many muffins were eaten? *Ans:* 11 muffins.

10. What's the difference between the brothers' ages if one is 20 years old and the other is only 12? *Ans:* 8 years.

11. Malik is 67 years old, which is 22 years older than his nephew. How old is his nephew? *Ans:* 45 years old.
    Solution: Malik is older than his nephew by 22 years. The nephew, who is younger than Malik, has 67 (years) - 22 (years) = 45 (years).

12. Ravina is 17 years old, which is 12 years younger than her teacher. How old is the teacher? *Ans:* 29 years.
    Solution: The teacher is older than Ravina by 12 years. The teacher is 17 (years) + 12 (years) = 29 (years).

13. How old is the son if the father is 46 years old and the son is 21 years younger? *Ans:* 25 years (46 - 21 = 25).

14. How old is the mom if the daughter is 27 and her mom is 21 years older? *Ans:* 48 years old.

15. Carly's job is to keep 50 feet of front walk clean.
    a) After she swept 12 feet, how many feet remain to be swept? *Ans:* 38 feet.
    b) After she swept 22 feet, how many feet remain to be swept? *Ans:* 28 feet.
    c) After she swept 32 feet, how many feet remain to be swept? *Ans:* 18 feet.

16. A man has 76 miles to travel, and on the first day he travels 22 miles. How many miles are left to travel? *Ans:* 54 miles.

17. After paying $60 for a dress, Hannah received $11 in change. How much was the dress? *Ans:* $49.

18. There were 55 boxes in storage and Hank took out 22 of them. How many boxes were left in the storage? *Ans:* 33 boxes.

19. There were 57 flies in the room and 21 more flew in. How many flies are in the room now? *Ans:* 78 flies. That must be awful! Maybe they are butterflies.

20. 57 worker ants were carrying a stick. Then 12 ants quit. How many worker ants kept on with the stick? *Ans:* 45 ants.

21. How much heavier are 48 pounds of potatoes than 11 pounds of onions? *Ans:* 37 pounds.

22. How much more expensive is a $63 coat than a $12 sweatshirt? *Ans:* $51.

23. How much longer is a 41-foot boat than a 21-foot boat? *Ans:* 20 feet.

24. How much shorter is a 41-foot boat than a 51-foot boat? *Ans:* 10 feet.

25. How much younger is a 21 year old than a 32 year old? *Ans:* 11 years.

26. How many people will remain on a 58 passenger bus after 23 passengers get off? *Ans:* 35 passengers.

27. A geologist found 26 pieces of gold but 22 pieces were "fool's gold". How many pieces of real gold did she find? *Ans:* 4 pieces of real gold.

28. A swimming coach worked for 50 years. For 32 years, she worked with the school team, and then she coached the Olympic teams. How many years did she coach the Olympic team? *Ans:* 18 years.

29. A teapot and a teacup together cost $48. What's the price of the teapot if the teacup is $12? *Ans:* $36.

30. A locksmith sold 82 items. Of these, 31 were locks and the rest were keys. How many keys did he sell? *Ans:* 51 keys.

31. A crocodile with 77 sharp teeth used 41 of them to bite Captain Hook. How many of his teeth missed Captain Hook's wooden leg? *Ans:* 36 teeth.

32. If a swimmer was scheduled to go to 30 swim meets but only went to 11, how many swim meets did she miss? *Ans:* 19 swim meets.

33. If Leo cut 21 feet from a 50-foot wire, how much wire remains? *Ans:* 29 feet.

34. From a 60 liters barrel a horse drank 31 liters. How many liters are still in the barrel? *Ans:* 29 liters. Did you know that a healthy horse can drink from 30 to 50 liters of water a day?

35. There were 40 monkeys in a bag and Gill let 21 of them out of the bag. How many monkeys are in the bag now? *Ans:* 19 monkeys.

36. Cyrus knows 100 knock-knock jokes. He told us 11 jokes before the bell rang. How many more jokes does he have left to tell? *Ans:* 89 jokes.

37. Sarah invited 20 friends for her doll's birthday party, but only 12 friends came. How many didn't come? *Ans:* 8 friends.

38. Five dozen eggs equals 60 eggs. If you take away 1 dozen, how many eggs will be left? *Ans:* 4 dozen or 48 eggs (60 - 12 = 48).

39. A car rental shop had 60 cars and rented out 22 of them. How many cars are not rented? *Ans:* 38 cars.

40. Michael borrowed $70 from Neil and returned $42. How much more does he owe? *Ans:* $28.

41. Terry has to take 60 units in college and has already taken 32. How many units does she still need to take? *Ans:* 28 units.

42. A school principal ordered 70 desks and received 52. How many desks will be coming? *Ans:* 18 desks.

43. Prudence's grandmother is coming in 50 days. If Prudence has already been waiting for 22 days, how many more days does she have to wait until her grandmother arrives? *Ans:* 28 days more.

44. If there are 50 problems to be done and you have already completed 39, how many do you have left before you're finished? *Ans:* 11 problems.

45. There are 61 lines in the song. If the singer has already sung 22 lines, how many are left to sing? *Ans:* 39 lines.

46. The veterinarian has to do 40 checkups today. By lunchtime, he has already done 22. How many checkups are left to do? *Ans:* 18 checkups.

47. Marty needs to send 39 letters today. He gave 12 to Susan to send. How many will he mail himself? *Ans:* 27 letters.

48. Ali has to peel all 52 of his vegetables before his mother comes home. If he has already peeled 11 eggplants and 12 carrots, how many more vegetables does he have to peel?
Ans: 29 vegetables.

49. Chloe and Ross collect stamps. If Chloe has 51 stamps in her collection and Ross has 21 stamps less than Chloe, how many stamps do they have together? *Ans:* 81 stamps.

50. There were 30 children in the class. 11 children wore red sweaters and 11 wore gray sweater. The rest wore blue sweaters. How many children are wearing blue sweaters?
Ans: 8 children.

ADDING NUMBERS ENDING IN 3 OR 4

EXERCISE I

20 + 13 = 33	33 + 16 = 49	44 + 44 = 88	53 + 27 = 80
21 + 13 = 34	45 + 24 = 69	24 + 65 = 89	23 + 24 = 47
22 + 13 = 35	64 + 33 = 97	25 + 65 = 90	67 - 32 = 35
32 + 23 = 55	43 + 35 = 78	23 + 17 = 40	87 - 51 = 36
63 + 13 = 76	36 + 33 = 69	37 + 23 = 60	76 - 22 = 54
24 + 12 = 36	43 + 55 = 98	36 + 43 = 79	77 - 22 = 55
24 + 14 = 38	44 + 55 = 99	44 + 26 = 70	58 - 41 = 17
53 + 24 = 77	33 + 44 = 77	56 + 34 = 90	88 - 32 = 56
35 + 42 = 77	22 + 55 = 77	23 + 57 = 80	94 - 62 = 32
26 + 53 = 79	55 + 33 = 88	16 + 64 = 80	98 - 61 = 37
14 + 14 = 28	33 + 33 = 66	26 + 34 = 60	49 - 21 = 28

WORD PROBLEMS

1. There are 40 days until the winter Olympics and then there are
    13 days of the Olympics. How long is it from now until the end
    of the Olympics? *Ans:* 53 days.

2. Harry has $31 in the bank.
   a) How much will it be if he puts in $13 more? *Ans:* $44.
   b) How much will it be if he puts in $23 more? *Ans:* $54.
   c) How much will it be if he puts in $33 more? *Ans:* $64.

3. A journey took 23 miles across a valley and 11 miles through a canyon. How long was the journey? *Ans:* 34 miles. Journey means a trip.

4. A store was open 31 days last month and 23 days this month. How many days was the store open? *Ans:* 54 days.

5. A plumber connected a 42-inch pipe with a 33-inch pipe. How long are the two pipes together? *Ans:* 75 inches.

6. If a music teacher tuned 14 guitars and 13 violins, how many musical instruments did he tune? *Ans:* 27 instruments.

7. A car wash washed 34 small cars and 23 trucks. How many vehicles did it wash? *Ans:* 57 vehicles.

8. A frog jumped 53 inches and then 34 inches more. How many inches did it jump in the two jumps? *Ans:* 87 inches.

9. One stamp is worth 35 cents, another 23 cents. How much are both stamps worth? *Ans:* 58 cents.

10. A trader swapped 23 picture frames for 45 frying pans. How many items were swapped? *Ans:* 68 items.

11. If Josh earned $45 the first day and $43 the second, how much did he earn in two days? *Ans:* $88.

12. Carl and Liz started a Pet Loving Club. Carl signed up 23 kids and Liz signed up 24. How many children are in the PL club? *Ans:* 49 students. Did you forget Carl and Liz?

13. At a tennis game, 43 fans rooted for the home team and 42 cheered for the guests. How many fans were at the game? *Ans:* 85 fans.

14. After Neil puts 55 beads in a jar that already had 33 beads in it, how many beads will be in the jar? *Ans:* 88 beads.

15. For the school picnic, we made 26 turkey sandwiches and 34 sandwiches with tuna. How many sandwiches did we make? *Ans:* 60 sandwiches.

16. A backpack cost $47 and the sleeping bag cost $43. How much did they both cost? **Ans:** $90.

17. On the lake, Marsha saw 54 ducks and 26 swans. How many birds did Marsha see on the lake? **Ans:** 80 birds.

18. How many total workers are 45 carpenters and 23 locksmiths together? **Ans:** 68 workers.

19. How many books will it be if we put 64 novels and 15 mysteries together? **Ans:** 79 books.

20. An art teacher collected 73 drawings and 17 paintings. How many works of art did the teacher collect?
    Ans: 90 works of art.

21. An optician sold 18 prescription glasses and 13 sunglasses. How many glasses did he sell altogether? **Ans:** 31 glasses.

22. One bag weighs 38 pounds and the other weighs 33 pounds. How much do they both weigh? **Ans:** 71 pounds.

23. In the morning Julian counted 36 roses and 33 iris.
    a) How many flowers did Julian count? **Ans:** 69 flowers.
    b) After he counted again he realized that there were actually 34 iris. How many flowers did he count the second time?
    Ans: 70 flowers.

24. A royal jeweler used 44 diamonds and 25 rubies for the King's new crown. How many precious stones did the jeweler use?
    Ans: 69 stones.

25. For the Donate-a-Toy Day we collected 34 toys and the neighbors brought in 44 toys. How many toys were gathered altogether? **Ans:** 78 toys.

26. One cow drank 19 gallons of water, the other drank 23 gallons. How many gallons did both cows drink? **Ans:** 42 gallons.

27. A cow gave 43 glasses of milk in the morning and 38 glasses in the evening. How many glasses of milk did it give a day?
    Ans: 81 glasses.

28. Two men traveled in opposite directions. One went 29 miles and the other walked 33 miles. How many miles are between them? **Ans:** 62 miles.

29. After Mr. Hardy built 37 feet of the fence, he still had 24 more feet to build. How long will the fence be when it's finished? *Ans:* 61 feet.

30. My digital clock shows 23 minutes after the hour. The clock is 18 minutes behind. What is the time? *Ans:* 41 minutes after the hour.

31. A sales clerk divided the purchases into two bags, 26 items in one bag and 14 in the other. How many items were in both bags? *Ans:* 40 items.

32. A kiosk sold 56 newspapers and 14 magazines. How many periodicals did it sell? *Ans:* 70 periodicals.

33. During an archery competition,
    a) Lynn shot 26 arrows into the target and missed 14 times. How many arrows did she shoot? *Ans:* 40 arrows.
    b) Maggie shot 27 arrows into the target and missed 13 times. How many arrows did she shoot? *Ans:* 40 arrows.
    c) Nadia shot 28 arrows into the target and missed 12 times. How many arrows did she shoot? *Ans:* 40 arrows.

34. Mrs. Peters sent two packages, 34 pounds each. What was the weight of both packages? *Ans:* 68 pounds.

35. Lorna picked up 2 magazines, one with 37 pages of ads and the other with 44 pages of ads. How many pages of ads were in both magazines? *Ans:* 81 pages.

36. A baker sold 24 poppy seed bagels and 48 plain bagels. How many bagels were sold? *Ans:* 72 bagels.

37. Another baker sold 24 sesame seed bagels and 39 onion bagels. How many bagels did she sell? *Ans:* 63 bagels.

38. Kira divided a bag of cherries into two plates, 28 cherries in one and 33 cherries in the other. How many cherries were in the bag? *Ans:* 61 cherries.

39. Nan and Don were blowing bubbles. Nan blew 39 bubbles and Don blew 43. How many bubbles did they blow together? *Ans:* 82 bubbles.

40. A lady in a hat store tried on 24 summer hats and 27 spring hats. How many hats did she try? *Ans:* 51 hats.

41. A man in a tie store tried on 14 silk ties and 29 satin ties. How many ties did the man try on? *Ans:* 43 ties.

42. A dog chased 38 tabby cats and 33 Persian cats. How many cats did it chase? *Ans:* 71 cats.

43. When a motorcycle racer fell the first time he needed 18 stitches. The second time it took 14 more stitches than the time before. How many stitches did the racer need the second time? *Ans:* 32 stitches.

44. A zookeeper takes care of 29 reptiles and 24 birds. How many animals are under his care? *Ans:* 53 animals.

45. During the football season one team scored 39 touchdowns and the other team scored 54. How many touchdowns did both teams score? *Ans:* 93 touchdowns.

46. From the ground floor there are 27 steps down to the basement and 33 steps up to the second floor. How many steps are from the second floor to the basement? *Ans:* 60 steps.

47. Mr. Hedge is building a fence.
a) If he bought 37 short posts and 24 long ones, how many posts did he buy? *Ans:* 61 posts.
b) If he cut a rope into two pieces, 28 feet and 34 feet, how long was the rope before he cut it? *Ans:* 62 feet.
c) If he used 49 long nails and 33 short ones, how many nails did he use? *Ans:* 82 nails.

48. Amy and Bella are jumping rope. Amy jumped 67 times and Bella jumped 24. How many jumps did they make altogether? *Ans:* 91 jumps.

49. During a snowball fight, we threw 36 snowballs and they threw 43. How many snowballs were thrown in the fight? *Ans:* 79 snowballs.

50. A monkey's body is 59 centimeter; its tail is 22 centimeter. How long is the monkey including its tail? *Ans:* 81 centimeter.

SUBTRACTING NUMBERS
ENDING IN 3 OR 4

EXERCISE I

24 + 13 = 37	54 + 42 = 96	43 + 35 = 78	55 + 33 = 88
23 + 23 = 46	34 + 53 = 87	63 + 33 = 96	23 + 33 = 56
33 + 13 = 46	74 + 14 = 88	43 + 55 = 98	44 + 45 = 89
44 + 12 = 56	63 + 16 = 79	44 + 55 = 99	24 + 65 = 89
24 + 14 = 38	44 + 24 = 68	33 + 44 = 77	35 + 65 = 100
53 + 24 = 77	64 + 33 = 97	22 + 55 = 77	35 + 35 = 70

EXERCISE II

23 - 14 = 9	26 + 14 = 40	58 - 41 = 17	30 - 26 = 4
23 + 14 = 37	53 - 27 = 26	58 + 41 = 99	17 - 4 = 13
36 - 23 = 13	53 - 24 = 29	33 - 13 = 20	17 + 4 = 21
36 + 23 = 59	27 - 14 = 13	33 + 13 = 46	27 + 5 = 32
56 - 34 = 22	27 + 14 = 41	41 - 21 = 20	13 + 13 = 26
56 + 34 = 90	34 - 24 = 10	41 + 21 = 62	14 + 13 = 27
26 - 14 = 12	34 + 24 = 58	30 - 24 = 6	35 - 13 = 22

WORD PROBLEMS

1. Carol planted 44 carrot seeds but only 33 carrots sprouted. How many seeds didn't come up? *Ans:* 11 seeds.

2. Bob paid for cat food with a $50 bill and got back $43 in change. How much was the cat food? *Ans:* $7.

3. Jay printed 53 copies of his collection of poems and gave 23 copies to his friends. How many copies were left? *Ans:* 30 copies.

4. There are 24 rides in a theme park and Julian rode 13. How many more are left for him to ride? *Ans:* 11 more rides.

5. A truck that carried 55 boxes delivered 43 boxes to the first address. How many boxes are inside the truck now? *Ans:* 12 boxes.

6. 46 seagulls were sitting on the beach before we came. Then 23 seagulls flew to the right and the rest took off to the left. How many flew to the left? *Ans:* 23 seagulls.

7. On a 54 day cruise, the cruise ship spent 24 days in ports and the rest on ocean. How many days was on the ship on the ocean? *Ans:* 30 days.

8. Out of 65 bread pieces Hansel and Gretel dropped on the path, birds ate 53. How many bread pieces were left on the path? *Ans:* 12 pieces.

9. If out of 53 sunflower seeds, blue jays ate 33, how many seeds were left for the finches? *Ans:* 20 seeds.

10. How many books were left on the bookshelf if there were 65 books before Keith took 43 books off the shelf? *Ans:* 22 books.

11. From a 45 pound-bag of flour, a baker took 14 pounds. How many pounds of flour did he leave in the bag? *Ans:* 31 pounds.

12. One story is 44 pages long; the other story is 24 pages longer. How long is the second story? *Ans:* 68 pages.

13. Out of 37 acres of land, a farmer will plant corn on 33 acres and leave the rest empty. How many acres of empty land will there be? *Ans:* 4 acres.

14. The pine tree is 76 feet tall. The birch tree is 34 feet shorter. How tall is the birch tree? *Ans:* 42 feet.

15. A high school has 77 teachers and 14 other helpers. How many people are working at the school? *Ans:* 91 people.

16. At the cabin we found 56 firewood logs and burned 13 the first day. How many logs were left for the rest of our stay? *Ans:* 43 logs.

17. Tyler knows 39 kids whose name is Alex. If there are 13 girls with that name, how many boys are named Alex? *Ans:* 26 boys.

18. Alex knows 29 kids whose name is Tyler. If there are 14 girls with that name, how many boys are named Tyler? *Ans:* 15 boys.

19. There are 46 bolts and 23 nuts in a box. How many more bolts than nuts are there? *Ans:* 23 more bolts.

20. Last year Conrad could run around the block in 75 seconds. This year he runs faster by 13 seconds. How fast can he run this year? *Ans:* 62 seconds.

21. Don had to travel 75 miles and now has only 23 more miles to go. How many miles has he already covered? *Ans:* 52 miles.

22. Luanne keeps 83 stamps in two envelopes with 44 stamps in the first envelope. How many stamps are in the second envelope? *Ans:* 39 stamps.

23. Hannah cut 53 inches from a 67-inch tape. How long is the remaining tape? *Ans:* 14 inches.

24. The earrings and the necklace cost $38. The necklace is $14. How much are the earrings? *Ans:* $24.

25. My teacher gave me 45 days to finish the project but I was done in 33. By how many days was I ahead of schedule? *Ans:* by 12 days.

26. Kim was paid $43 and Heather was paid $68. How much more did Heather receive? *Ans:* $25 more.

27. There are 58 apple trees in the orchard. There are 24 fewer cherry trees than apple trees. How many cherry trees are in the orchard? *Ans:* 34 cherry trees.

28. A store bought 49 bags of tomatoes and sold 14 bags the first day. How many bags were left? *Ans:* 35 bags.

29. A small theater expected 54 people for their first show but 67 people came. How many people came who were not expected? *Ans:* 13 people.

30. Out of 88 keys in the piano, 44 were out of tune. How many were tuned? *Ans:* 44 keys.

31. Out of 53 errors Eric made on his math test, 23 were in addition, the rest in subtraction. How many errors did Eric make in subtraction? *Ans:* 30 errors.

32. One mole dug a 42-foot tunnel. The second mole's tunnel is 3 feet shorter. How long is the tunnel dug by the second mole? *Ans:* 39 feet.

33. In a county competition, the first-prize pumpkin weighed 40 pounds. The second place pumpkin was 13 pounds lighter. How much did the second pumpkin weigh? *Ans:* 27 pounds.

34. The archeologists found 50 ancient vases, 23 of them broken. How many unbroken vases did they find? *Ans:* 27 vases.

35. Ricardo had $60 and spent $34 on a watch. How much money does he have now? *Ans:* $26.

36. Two boys picked 52 apples. One picked 23. How many apples did the second boy pick? *Ans:* 29 apples.

37. There are 22 frozen pizzas in the freezer and 13 frozen corn dogs. How many more pizzas than corn dogs are in the freezer? *Ans:* 9 pizzas.

38. There are 43 airplanes and 14 helicopters at an airport. How many more airplanes then helicopters are there? *Ans:* 29 more airplanes.

39. Ross and Sam together found 82 snails. If Ross found 43, how many snails did Sam find? *Ans:* 39 snails.

40. One of Hunter's dogs weighs 62 pounds, the other weighs 53. What's the difference in their weights? *Ans:* 9 pounds.

41. Jerry picked 43 tomatoes from his garden. If he gave 14 tomatoes to the neighbor, how many did he keep? *Ans:* 29 tomatoes.

42. A shop bought a broken bicycle for $24, then fixed and sold it for $63. How much money did the shop make? *Ans:* $39.

43. A birch tree is 23 feet shorter than a pine tree. How tall is the birch if the pine tree is 71 feet tall? *Ans:* 48 feet.

44. Sylvia weighed 72 pounds and gained 24 more pounds. How much does she weigh now? *Ans:* 96 pounds.

45. The father is 33 years older than the son. If the father is 52 years old, how old is the son? *Ans:* 19 years old.

46. The daughter is 23 years younger than her mother. If the mother is 61 years old, how old is the daughter? *Ans:* 38 years old.

47. The grandpa is 64 years older than his grandson. If the grandpa is 71 years old, how old is the grandson? *Ans:* 7 years old.

48. In one week Mr. Redwood built 34 feet of a 81-foot fence. How many more feet of the fence are left for him to build? *Ans:* 47 feet.

49. Nick and Ogden planted 62 trees. If Ogden planted 33 trees how many trees did Nick plant? *Ans:* 29 trees.

50. There are 51 olives in two jars. If there are 24 olives in one, how many are in the other? *Ans:* 27 olives.

51. At rest Garrett's heart rate was 64 beats per minute. After the exercise it went up to 91. By how much did his heart rate increase? *Ans:* 27 beats per minute. What is a heart rate? Do you know your heart rate?
Please show child how to count his heart beats on his wrist in a 30-second period. Then ask child to double that amount.

ADDING NUMBERS ENDING IN 9

EXERCISE I

33 - 9 = 24	77 - 29 = 48	13 + 9 = 22	44 + 9 = 53
53 - 9 = 44	58 - 29 = 29	14 + 9 = 23	64 + 9 = 73
73 - 9 = 64	88 - 29 = 59	15 + 9 = 24	43 + 19 = 62
56 - 9 = 47	98 - 39 = 59	25 + 9 = 34	45 + 19 = 64
76 - 9 = 67	93 - 39 = 54	35 + 9 = 44	47 + 19 = 66
38 - 9 = 29	73 - 49 = 24	45 + 9 = 54	49 + 19 = 68
48 - 9 = 39	53 - 49 = 4	55 + 9 = 64	59 + 19 = 78
53 - 19 = 34	78 - 59 = 19	59 + 5 = 64	69 + 9 = 78
23 - 19 = 4	88 - 69 = 19	55 + 9 = 64	55 + 9 = 64
33 - 19 = 14	88 - 79 = 9	65 + 9 = 74	33 + 9 = 42
53 - 19 = 34	98 - 89 = 9	74 + 9 = 83	44 + 9 = 53
73 - 29 = 44	88 - 19 = 69	84 + 9 = 93	24 + 9 = 33

WORD PROBLEMS

1. At a gas station, Del paid $15 for gasoline and $9 for a snack. How much did he spend? *Ans:* $24.

2. At the same gas station, Del paid $9 for a tire fix-it kit and $17 for oil. How much did he spend on these two items? *Ans:* $26.

3. It was 45 degrees in the morning and then the temperature went up 9 degrees. What is the temperature now? *Ans:* 54 degrees.

4. It was 41 degrees and then the temperature went down by 9 degrees. What is temperature now? *Ans:* 32 degrees.

5. Blair said 27 "Thanks you" and 9 "Excuse me" during a visit to his aunt. How many polite words did he use? *Ans:* 36 polite words.

6. Mom spent $43 at the grocery store and $9 at the pharmacy. How much money did she spend? *Ans:* $52.

7. Mom spent $13 for pastries and $9 for a special cake. How much did mom spend at the bakery? *Ans:* $22.

8. We went for a 15 mile bike ride yesterday, but today we did 19 miles. How many total miles did we go in two days?
   Ans: 34 miles. 23 - 9 = 14

9. On a 34-mile return trip, we biked 17 miles the first day. How many miles did we do the second day? *Ans:* 17 miles.

10. During a trip, Ali sent 19 post cards from New York City and 21 cards from Boston. How many cards did he send?
    Ans: 40 cards.

11. A jacket is $29; the running shoes are $12. How much do they both cost? *Ans:* $41.

12. Two pairs of jeans cost $19 each. How much will it cost to buy 2 of them? *Ans:* $38.

13. A pair of pants cost $19 and matching shirt is $23. How much will both cost? *Ans:* $42.

14. Brenda's airline ticket cost $59 which included $14 for the airport tax. How much did she pay for the trip without including the tax? *Ans:* $45.

15. At the airport Brenda spent $7 for a sandwich and $29 on a gift for her sister. How much did she spend at the airport?
    Ans: $36.

16. While waiting for her sister at the magazine stand, Brenda bought two magazines for $9 and 3 paperback books for $19. How much did she spend now? *Ans:* $28.

17. From the airport Brenda had to take a taxi. She paid $55 for the trip and also gave $9 tip. How much did she pay to the cab driver? *Ans:* $64.

18. The temperature in the morning was 53 degrees. By noon, it was up 19 degrees. What was the temperature at noon? *Ans:* 72 degrees.

19. A farmer, whose farm had 32 acres, bought 19 more acres of land. How big is the farm now? *Ans:* 51 acres.

20. The farmer has 51 acres now. He has two sons and he wants to give 19 acres to his first son. How many acres does he have left? *Ans:* 32 acres.

21. Darryl rode a bicycle 19 miles one day and 24 miles the second day. How many miles did he ride? *Ans:* 43 miles.

22. One classroom has 29 students and the other also has 29 students. How many students are in both classrooms? *Ans:* 58 students.

23. Ella got 29 points on the first test and 39 on the second test. How many total points did she get so far? *Ans:* 68 points.

24. In English, Jack got 29 points out of 30. In math, he got 15 points out of 20. How many total points did he get on both tests? *Ans:* 44 points.

25. Tony, the dog, sat for 19 minutes and then slept for 22 minutes waiting for his treat. How long did he wait? *Ans:* 41 minutes.

26. Tony, the dog, ran in circles to the right 34 times, then to the left 29 times. How many times did he run in both directions? *Ans:* 63 times.

27. Tony, the dog, chased 15 squirrels and 19 birds today. How many animals did he chase? *Ans:* 34 animals.

28. Tony, the dog, barked 37 times at passing cars and 39 times at pedestrians. How many times did he bark? *Ans:* 76 times.

29. In a pasture there are 57 sheep and 29 cows. How many farm animals are there? *Ans:* 86 farm animals.

30. The farm has 28 goats. Of these, 19 are black goats and the rest are white. How many white goats are there? *Ans:* 9 goats.

31. A highway police officer gave a driver a speeding ticket for driving 19 kph an hour faster than the 65 kph speed limit. How fast did the driver go? *Ans:* 84 kph.

32. One electrician worked 28 hours; the other electrician worked 29 hours on the construction job. How many hours did they both work? *Ans:* 57 hours.

33. The carpenter worked 12 hours on the job and the plumber worked 19 hours. How many hours did they both work? *Ans:* 31 hours.

34. One downtown building cost $9 million to build. The second will cost $19 million more than that. How much will the second building cost? *Ans:* $28 million.

35. A movie producer hired 19 actors and 19 actresses for a new movie. How many people did she hire? *Ans:* 38 people.

36. The movie director hired 30 stage hands and 9 cameramen. How many people did he hire? *Ans:* 39 people. What does a director do?

37. At age 49, Mr. Dill went to the Fresh Herbs Company and worked there for 19 years before retirement. How old was Mr. Dill when he retired? *Ans:* 68 years old.

38. A grocery store sold 48 pounds of peaches and 48 pounds of plums. How many pounds of both fruits did it sell? *Ans:* 96 pounds. Solution: Let's just pretend it is 50 instead of 48. So we add 50 + 50 and get 100. But 50 is 2 larger than 48, so we subtract 4 from 100 and the answer is 96.

39. A rabbit running away from a wolf saw that he is 47 feet behind. So the rabbit began to run faster and increased the distance by 39 feet. How far behind is the wolf now? *Ans:* 86 feet.

40. On a slice of bread which weighs 44 grams, I spread 19 grams of jam. How much does my bread with jam weigh?
Ans: 63 grams.

41. At the station to a 59-car train they attached 14 more cars. How many cars does the train have? **Ans:** 73 cars.

42. The train could go as fast as 59 kph but will be going 19 kph slower. How much was its speed on the trip? **Ans:** 40 kph.

43. The fish head weighs 27 pounds; the rest of the fish weighs 69 pounds. How much does the whole fish weigh?
Ans: 96 pounds.

44. For the Fourth of July celebration, I bought 19 red balloons, 19 white balloons, and 19 blue balloons. How many balloons did I buy? **Ans:** 57 balloons.
Solution: 19 red balloons + 19 white balloons = 38 balloons. Then, 38 balloons + 19 blue balloons = 57 balloons.

45. A fish's tail weighs 19 pounds, its body is 39 pounds, and the head is 29 pounds. How much does the whole fish weigh?
Ans: 87 pounds.
Solution: First, let's find the weight of the tail and the body together: 19 lb + 39 lb = 58 pounds. Then, we will add the head: 58 lb + 29 lb = 87 lb.

46. Sally had 24 pears and gave Benny 15 of them. How many pears does Sally have now? **Ans:** 9 pears.

47. James has 27 melons in his vegetable patch. He threw away 8 because the birds had pecked them and then picked 9 good ones to take home. How melons are still on the vine?
Ans: 10 melons.

48. A woman bought 36 plants and planted 8 of them in pots. Of the remaining she planted 9 in the front of the house and the rest she planted in the back of the house. How many plants did she plant in the back of the house? **Ans:** 19 plants.

49. A girl bought four baskets of fruits. The strawberries cost $9, the peaches cost $13 and the apples cost $12. How much did she pay for the pears, if all baskets cost $43? **Ans:** $9.

50. We have three puppies that weigh 9 pounds, 13 pounds and 14 pounds. How much all the puppies weigh together? *Ans:* 36 pounds.

51. A birch tree is 27 feet tall. Its roots go 19 feet underground. How tall is it from top to the bottom of the roots? *Ans:* 46 feet.

SUBTRACTING NUMBERS ENDING IN 9

EXERCISE I

22 - 9 = 13	23 - 9 = 14	37 - 19 = 18	33 - 19 = 14
31 - 9 = 22	28 - 9 = 19	34 - 19 = 15	43 - 29 = 14
34 - 9 = 25	29 - 9 = 20	44 - 19 = 25	55 - 29 = 26
56 - 9 = 47	39 - 9 = 30	42 - 19 = 23	62 - 29 = 33
23 - 9 = 14	33 - 9 = 24	21 - 19 = 2	55 - 29 = 26
44 - 9 = 35	47 - 9 = 38	33 - 19 = 14	78 - 29 = 49
34 - 9 = 25	44 - 19 = 25	21 - 19 = 2	67 - 29 = 38
67 - 9 = 58	25 - 19 = 6	51 - 19 = 32	57 - 19 = 38

- To which number do you add 9 to make 18? *Ans:* 9
- To which number do you add 9 to make 30? *Ans:* 21
- To which number do you add 9 to make 37? *Ans:* 28
- To which number do you add 9 to make 46? *Ans:* 37
- To which number do you add 9 to make 88? *Ans:* 79
- To which number do you add 9 to make 55? *Ans:* 46
- To which number do you add 9 to make 64? *Ans:* 55
- To which number do you add 9 to make 75? *Ans:* 66

- To which number do you add 9 to make 54? *Ans:* 45
- To which number do you add 9 to make 33? *Ans:* 24
- To which number do you add 9 to make 62? *Ans:* 53
- To which number do you add 9 to make 41? *Ans:* 32
- To which number do you add 9 to make 81? *Ans:* 72
- To which number do you add 9 to make 82? *Ans:* 73

EXERCISE II

87 - 19 = 68	67 - 19 = 48	51 - 19 = 32	73 - 29 = 44
88 - 19 = 69	53 - 19 = 34	41 - 29 = 12	67 - 29 = 38
89 - 19 = 70	62 - 19 = 43	62 - 29 = 33	52 - 19 = 33
67 - 19 = 48	61 - 19 = 42	33 - 29 = 4	45 - 19 = 26
68 - 19 = 49	61 - 29 = 32	83 - 29 = 54	43 - 19 = 24

WORD PROBLEMS

1. At the station, to a 59-car train they attached 14 more cars. How many cars are there now? *Ans:* 73 cars.

2. The train could go as fast as $58 kph but will be going $19 kph slower. How much is its speed on the trip. *Ans:* 39 kph.

3. Michelangelo's statue of David in Florence is 19 feet tall with the pedestal (stand). If the pedestal is 6 feet, how tall is the statue? *Ans:* 13 feet.

4. On page 26, there are 39 problems. I did 19 of them. How many more problems are left for me to do? *Ans:* 20 problems. Don't get confused by the page number.

5. Sheila took a test that had 40 questions and answered 29 of them correctly. How many questions did she answer wrong? *Ans:* 11 questions.

6. Priscilla dug out 60 potatoes and put them in two baskets. How many potatoes are in the second basket if the first has 29? *Ans:* 31 potatoes.

7. My brother and I take turns for setting the dinner table. How many times did my brother set the table if I did it 19 times in a 30 days? *Ans:* 11 times.

8. The temperature outside is 70 degrees. The temperature in the pool is 49 degrees. By how much is the outside temperature warmer than the pool? *Ans:* 21 degrees.

9. In his closet Mr. Starch has 46 shirts; 19 shirts are white, the rest are colored shirts. How many colored shirts does he have? *Ans:* 27 shirts.

10. In her closet Mrs. Starch has 37 skirts; 29 skirts are long, the rest are short. How many short skirts does she have? *Ans:* 8 short skirts.

11. Johnny Starch has in his closet 15 pants and 9 shorts. How many shorts and pants are in his closet? *Ans:* 24 shorts and pants.

12. Mimi Starch has in her closet 52 diapers; 19 diapers are made of fabric, the rest are disposable. How many disposable diapers does Mimi have? *Ans:* 33 diapers.

13. The whole concert lasted 75 minutes. The intermission was 19 minutes. How long was the concert not counting intermission? *Ans:* 56 minutes.

14. There were 86 guests at the wedding; 39 from the bride's side and the rest were from the groom's guests. How many guests came from the groom's side? *Ans:* 47 guests.

15. Out of 86 guests at the wedding, 29 came from out of town. How many guests were local? *Ans:* 57 guests.

16. Among 86 guests, there were 49 women. How many men came to the wedding? *Ans:* 37 men.

17. Out of 86 guests, 59 ordered chicken for dinner. How many did not order chicken? *Ans:* 27 guests.

18. Boris borrowed $62 from Fred. He gave back $39 and said that he still owed $32. Did Boris count correctly? *Ans:* No. He owes $23, not $32..

19. Two friends bought a boat for $82. If one friend paid $39, how much did the other friend pay? *Ans:* $43.

20. If my uncle is 9 years older than my aunt and my aunt is 41, how old is he? *Ans:* 50 years old.

21. 21. If my cousin is 9 years younger than her husband, who is 38 years old, how old is she? *Ans:* 29 years old.

22. Mike's sister is 9 years older than Mike. How old is he if the sister is 22 years old? *Ans:* 13 years old.

23. If the father is 46 years old and is 39 years older than his son, how old is the son? *Ans:* 7 years old.

24. If a father is 46 years old and 29 years older than his daughter, how old is the daughter? *Ans:* 17 years old.

25. An artist sold his painting for $75. How much did he earn if he spent $29 on canvas and frame? *Ans:* $46.

26. In an office, only 29 feet of a 55-foot long hallway are covered with carpet. How many feet of the hallway are not covered? *Ans:* 26 feet.

27. There are 87 feet from the house to the main pipe. A plumber brought only 59 feet of connecting pipe. How many more feet of pipe does he need? *Ans:* 28 feet.

28. My dad holds 58 paperclips in both hands. If there are 29 paperclips in one hand, how many are in the other? *Ans:* 29 paperclips.

29. A DVD set cost $59. Brooke has a $25 gift card. How much more does she need to buy the set? *Ans:* $34.

30. There were 100 paperclips in the box and Tully used 19. How many are left in the box? *Ans:* 81 paperclips.

31. The top score on a test is 100 points. Xavier's score is 39 points below the top. What is his score? *Ans:* 61 points.

32. A military company has 100 soldiers and is made up of 2 platoons. If one platoon has 59 soldiers, haw many are in the other? *Ans:* 41 soldiers.

33. A king's crown, made of gold and silver, weighs 36 ounces. If there are 19 ounces of silver, how much gold is in the crown? *Ans:* 17 ounces. How big is an ounce anyway?

34. A queen has 90 pairs of shoes. If 39 pairs are running shoes, how many regular shoes does she have? *Ans:* 51 pairs.

35. Of her 51 pairs of shoes, 19 pairs are pretty, but not comfortable. How many comfortable shoes does the queen have? *Ans:* 32 pairs.

36. A jester made 63 jokes about the king and the queen. If he made 39 jokes about the king, how many jokes did he make about the queen? *Ans:* 24 jokes.

37. A poet wrote 50 poems about the king and the queen. If he wrote 49 poems about the queen, how many did he write about the king? *Ans:* 1 poem.

38. A dry sponge weighs 38 grams. Soaked with water, it weighs 99 grams. How much water can the sponge soak up? *Ans:* 61 grams of water.

39. An empty bucket weighs 9 pounds. Filled with sand it weighs 81 pounds. How much does the sand weigh? *Ans:* 72 pounds.

40. The same bucket holds 65 pounds of water. Its weight is 9 pounds. How much does it weigh with the water? *Ans:* 74 pounds.

41. A boat with cargo weighs 93 tons. An empty boat weighs 59 tons. How much does the cargo weigh? *Ans:* 34 tons. How big is a ton?

42. The farm has 61 birds 39 are chickens and the rest are turkeys. How many turkeys are on the farm? *Ans:* 22 turkeys.

43. One team scored 59 points, the other team scored 77. How many points separate the two teams? *Ans:* 18 points.

44. Harry took hairs from a bear and a hare. Out of 63 hairs, 29 were from the bear. How many hairs came from the hare? *Ans:* 34 hairs.

45. A hair comb cost 49 cents, a toothbrush cost 68 cents. What's the difference in price between the two? *Ans:* 19 cents.

46. A toothbrush cost 68 cents and the toothpaste cost 59 cents. What's the price difference between the two? *Ans:* 9 cents. There were 80 toothpicks in a box and Robert used 39. How many toothpicks are in the box now? *Ans:* 41 toothpicks.

47. Adam and Ben together put 62 apples in a basket. Ben put in 29 apples. How many did Adam put in the basket?
Ans: 33 apples.

48. In a century (100 years) the kingdom had drought for 19 years. How many years was there no drought? *Ans:* 81 years.

49. In a century, the kingdom was at war for 39 years. How many peaceful years did it enjoy? *Ans:* 61 years.

50. King Wise II ruled the kingdom for the first 49 years of the century. The rest of the time the kingdom was ruled by his son Dumb I. How long did Dumb I rule? *Ans:* 51 years. Too long!

51. In 100 years, the kingdom held 80 summer carnivals and 29 winter carnivals. How many more summer carnivals did the kingdom have? *Ans:* 51 more.

52. Grandpa Moe has 52 grandchildren. Grandpa Larry has 39. How many more grandchildren does grandpa Moe have?
Ans: 13 more grandchildren.

53. If Christopher made 50 wishes and only 29 came true, how many more of his wishes are left to come true? *Ans:* 21 wishes.

54. Adam and Ben together gathered 42 shiny pebbles at the beach. Ben found 29 pebbles. How many did Adam find?
Ans: 13 pebbles.

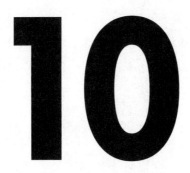

ADDING DOUBLE-DIGIT NUMBERS ENDING IN 5 OR 6

EXERCISE I

20 + 25 = 45	25 + 15 = 40	75 + 15 = 90	15 + 85 = 100
60 + 25 = 85	15 + 25 = 40	15 + 75 = 90	35 + 65 = 100
65 + 25 = 90	25 + 45 = 70	55 + 45 = 100	45 + 45 = 90

EXERCISE II

- To which number do you add 5 to make 18? *Ans:* 13
- To which number do you add 15 to make 30? *Ans:* 15
- To which number do you add 5 to make 37? *Ans:* 32
- To which number do you add 25 to make 46? *Ans:* 21
- To which number do you add 15 to make 88? *Ans:* 73
- To which number do you add 25 to make 55? *Ans:* 30
- To which number do you add 25 to make 64? *Ans:* 39
- To which number do you add 15 to make 75? *Ans:* 60
- To which number do you add 5 to make 54? *Ans:* 49

- To which number do you add 15 to make 33? *Ans:* 18
- To which number do you add 20 to make 62? *Ans:* 42
- To which number do you add 15 to make 41? *Ans:* 26
- To which number do you add 45 to make 81? *Ans:* 36
- To which number do you add 55 to make 80? *Ans:* 2

EXERCISE III

22 - 10 = 12	25 + 35 = 60	35 + 25 = 60	31 + 55 = 86
31 + 5 = 36	48 - 35 = 13	67 - 45 = 22	75 - 65 = 10
34 + 15 = 49	56 - 35 = 21	68 - 25 = 43	35 + 45 = 80
56 + 15 = 71	48 - 27 = 21	67 + 25 = 92	55 - 25 = 30
23 + 15 = 38	85 - 45 = 40	53 - 35 = 18	45 + 25 = 70
44 + 25 = 69	47 - 35 = 12	62 + 35 = 97	55 + 25 = 80
55 - 25 = 30	35 + 25 = 60	62 - 45 = 17	65 - 15 = 50
55 - 15 = 40	45 + 35 = 80	76 - 55 = 21	45 + 9 = 54

WORD PROBLEMS

1. Jill and Josh are playing table tennis. Jill has 15 points and Josh has 20 points. How many points do they have together? *Ans:* 35 points.

2. Miriam painted 35 flowers, Nita painted 25. How many flowers did they both paint? *Ans:* 60 flowers.

3. Paul sent 25 letters and 25 packages. How many pieces of mail did he send? *Ans:* 50 pieces of mail.

4. At the post office there were 45 men and 35 women. How many people were at the post office? *Ans:* 80 people.

5. A class period lasts 45 minutes and the break is 15 minutes. How long do they both last? *Ans:* 60 minutes or 1 hour.

6. There are 55 sharpened and 35 dull pencils in a drawer. How many pencils are there? *Ans:* 90 pencils.

7. After a 35 minute run, Jennifer walked for 25 minutes. How long was the run and walk combined? *Ans:* 60 minutes or 1 hour.

8. To 15 already on a rack, Heather added 35 dresses. How many dresses are on the rack now? *Ans:* 50 dresses.

9. A theater ticket costs $25 and a concert ticket is $35. What's the price of the two tickets? *Ans:* $60.

10. If theater and concert tickets cost $60 and a baseball game ticket cost $35. How much do all three tickets cost? *Ans:* $95.

11. The toaster cost $20 and the blender is $25 more.
    a) What's the price of the blender? *Ans:* $45.
    b) What's the price of both? *Ans:* $65 ($20 for toaster + $45 for blender).

12. There are 35 rabbits in one cage. The second cage has 15 more rabbits than the first. a) How many rabbits are in the second cage? *Ans:* 50 rabbits. b) How many rabbits are in both cages? *Ans:* 85 rabbits (35 + 50).

13. A coyote weighs 25 pounds and a wolf is 45 pounds more than coyote. a) How much does the wolf weigh? *Ans:* 70 pounds. b) How much do they both weigh? *Ans:* 95 pounds.

14. Oscar, the mouse, stole 2 quarters (a quarter is 25 cents). How many cents did Oscar steal? *Ans:* 50 cents.

15. Cole waited for a train for 25 minutes and then 26 more minutes. How long did he wait? *Ans:* 51 minutes.

16. One jar has 32 ounces of lemonade, the other has 35 ounces. How much lemonade is in the two jars? *Ans:* 67 ounces.

17. A bookshop sold 24 plain and 45 lined notebooks. How many notebooks did the bookshop sell? *Ans:* 69 notebooks.

18. Omar, the school's wrestling champion, ate 46 muffins and 45 pancakes. How many muffins and pancakes did he eat? *Ans:* 91 muffins and pancakes. Can anyone eat that much?

19. On a field trip, 27 children sat in the first bus and 25 children in the second. How many children went on the field trip? *Ans:* 52 children.

20. An ax costs $18 and a rake is $25. How much do they both cost? *Ans:* $43.

21. A cat costs $47 and the bag is $15. How much is the cat with the bag? *Ans:* $62.

22. There are 67 stamps in one album and 15 stamps in the other. How many stamps are in both? *Ans:* 82 stamps.

23. If 25 children played in the park and 17 more joined them later, how many children played in the park? *Ans:* 42 children.

24. At a tennis competition, a team won 35 games and lost 38. How many games did it play? *Ans:* 73 games.

25. A fish bit off a hook and a sinker. The hook cost 39 cents, the sinker was 45 cents. How much money did the fisherman lose? *Ans:* 84 cents.

26. In one day, a shoe shiner shined 36 men's shoes and 22 women's shoes. How many shoes did he shine? *Ans:* 58 shoes.

27. Two chandeliers hang in the dance hall. One has 46 light bulbs and the other has 44. How many light bulbs are in both? *Ans:* 90 light bulbs.

28. One mosquito fish ate 34 mosquitoes, meanwhile another ate 36. How many mosquitoes did they eat together? *Ans:* 70 mosquitoes.

29. A squirrel found 37 pinecones and 46 acorns. How many cones and acorns altogether did it find? *Ans:* 83 pinecones and acorns.

30. The squirrel also found 27 mushrooms and 55 berries. How many mushrooms and berries did it find? *Ans:* 82 mushrooms and berries.

31. The first winter a family burned 47 logs, the second winter it burned 46 logs. How many logs did they burn in the 2 winters? *Ans:* 93 logs.

32. Trudy fed 25 ducks and 26 ducklings. How many birds did she feed? *Ans:* 51 birds.

33. Trudy also fed 36 pigs and 35 piglets. How many pigs and piglets did she feed? *Ans:* 71 pigs and piglets.

34. In a pond, Mr. Pike counted 45 koi fish and 46 goldfish. How many fish did Mr. Pike count? *Ans:* 91 fish.

35. In my garden, 16 roses and 14 poppies are blooming. How many flowers are blooming? *Ans:* 30 flowers.

36. At the entrance to the school we planted 16 trees and 26 bushes. How many plants did we plant? *Ans:* 42 plants.

37. After my grandpa blew out 44 candles on his birthday cake, 26 candles were still burning. How old is grandpa?
Ans: 70 years old.

38. There are 36 goldfish in one tank and 34 in the other. How many goldfish are in both tanks? *Ans:* 70 goldfish.

39. Wally used 54 feet of wallpaper for the family room and 46 feet for the bedroom. How much wallpaper did Wally use?
Ans: 100 feet.

40. Mary fed 45 sheep and 46 lambs. How many sheep and lamb did she feed? *Ans:* 91 sheep and lamb.

41. In a restaurant, 38 customers ordered tea and 41 had coffee. How many drinks did the restaurant serve? *Ans:* 79 drinks.

42. An electrician installed 34 switches and 35 electric sockets. How many switches and sockets did she install?
Ans: 69 switches and sockets.

43. At the base, there were 28 soldiers living in one barrack and 36 soldiers in the other. How many soldiers lived in both barracks?
Ans: 64 soldiers.

44. There were 64 soldiers and 26 officers at a camp. How many military men and women were at the camp? *Ans:* 90 people.

45. Mrs. McLean loaded a dishwasher with 26 forks and 29 spoons. How many pieces of silverware did she put in the dishwasher?
Ans: 55 pieces.

46. She also put 37 salad plates and 46 dinner plates in the dishwasher. How many plates were in the washer? *Ans:* 83 plates. Did she have a party?

47. There were 18 kitchen knives and 65 dinner knives to be washed. How many knives were to be washed? *Ans:* 83 knives.

48. At first 28 fire fighters came to a big fire. Then 45 more came. Altogether, how many fire fighters came to put out the fire? *Ans:* 73 fire fighters.

49. From a peach tree Lucy picked 49 ripe peaches and she picked 46 ripe plums from a plum tree. How many fruits did she pick from both trees? *Ans:* 95 fruits.

50. From a vegetable patch, Lucy picked 38 tomatoes and 36 carrots. How many vegetables did she pick? *Ans:* 74 vegetables.

51. From one bush, Lucy picked 28 raspberries and from another bush she picked 25 blackberries. How many berries did she pick? *Ans:* 53 berries.

52. Ms. Baker put 47 raisins in her muffins and kept 36 to bake the cookies. How many raisins did she have at the start? *Ans:* 83 raisins.

53. A famous movie actor played in 15 comedies, 26 dramas, and 25 adventure movies. How many movies did he play in? *Ans:* 66 movies. Remember to add 15 and 25 first.

54. In a carpenter's toolbox there are 35 drill bits, 27 pliers, and 25 screwdrivers. How many tools are there? *Ans:* 87 tools.

55. For her birthday Chelsea received $18 from her aunt, $25 from her dad, and $15 from her sister. How much money did she get? *Ans:* $58.

56. An architect designed a building with 35 windows on the 1st floor, 29 windows on the 2nd, and 25 on the 3rd. How many windows are on all three floors? *Ans:* 89 windows.

57. Inside a piñata, kids found 16 chocolate bars, 35 lollypops, and 35 caramels. How many candies were inside the piñata? *Ans:* 86 candies.

58. On the shore Marina saw 26 motor boats, 25 dinghies, 24 yachts, and 25 canoes. How many boats did Marina see? *Ans:* 100 boats.

Solution: Let's regroup the numbers. We will add 25 dinghies to 25 canoes (25 + 25 = 50). Then we add 26 motor boats and 24 yachts (26 + 24 = 50). Then, 50 + 50 = 100.

59. Kids learned about 15 types of conifer tree, 15 types of deciduous trees and 9 types of fruit trees. How many different trees did they learn about? *Ans:* 39 trees.
Conifers are evergreen trees like pines, deciduous are trees that drop their leaves in winter, like maples and most of the fruit trees; apples, peaches, apricots, plum, nectarine, persimmon, etc.

SUBTRACTING DOUBLE-DIGIT NUMBERS ENDING IN 5 OR 6

EXERCISE I

22 - 15 = 7	56 + 16 = 72	55 - 16 = 39	48 - 16 = 32
31 + 16 = 47	23 + 15 = 38	55 - 15 = 40	56 - 26 = 30
34 + 15 = 49	44 + 16 = 60	25 + 16 = 41	45 + 26 = 71

EXERCISE II

- To which number do you add 5 to make 14? *Ans:* 9
- To which number do you add 5 to make 30? *Ans:* 25
- To which number do you add 5 to make 37? *Ans:* 32
- To which number do you add 5 to make 46? *Ans:* 41
- To which number do you add 5 to make 81? *Ans:* 76
- To which number do you add 5 to make 49? *Ans:* 44
- To which number do you add 5 to make 64? *Ans:* 59
- To which number do you add 6 to make 69? *Ans:* 63
- To which number do you add 6 to make 48? *Ans:* 42

- To which number do you add 6 to make 27? *Ans:* 21
- To which number do you add 6 to make 56? *Ans:* 50
- To which number do you add 6 to make 35? *Ans:* 29
- To which number do you add 6 to make 75? *Ans:* 69
- To which number do you add 6 to make 76? *Ans:* 70

EXERCISE III

100 - 15 = 85	100 - 45 = 55	100 - 90 = 10	100 - 55 = 45
100 - 16 = 84	100 - 55 = 45	100 - 15 = 85	100 - 45 = 55
100 - 25 = 75	100 - 65 = 35	100 - 25 = 75	100 - 30 = 70
100 - 35 = 65	100 - 85 = 15	100 - 50 = 50	100 - 45 = 55

WORD PROBLEMS

1. Eight ducks and 7 geese together need 60 pounds of food. If the ducks need 25 pounds, how much food do the geese need? *Ans:* 35 pounds.

2. In the beginning of the year, 51 children signed up for the ballet class. By the end of the year only 35 children remained in the class. How many dropped out? *Ans:* 16 children.

3. The old swimming record was 55 seconds. The new record is 46 seconds. By how many seconds is the new record better? *Ans:* 9 seconds.

4. To make a cherry jam, Jared took 95 cherries and pitted 25 of them. How many cherries are left to be pitted? *Ans:* 70 cherries.

5. A writer finished 46 pages of his 90 page novel. How many more pages does he still have to write? *Ans:* 44 pages.

6. A professor is 76 years old; her assistant is 35 years younger. How old is the assistant? *Ans:* 41 years old.

7. It takes 38 minutes to walk to school. The bus gets there in 15 minutes. How much time do I save by taking the bus? *Ans:* 23 minutes.

8. Kermit and his backpack weigh 70 pounds. The backpack weighs 16 pounds. How much does Kermit weigh? *Ans:* 54 pounds.

9. A clown started the day with 57 balloons. By the end of the day he had only 26 balloons left. How many balloons did he give away? *Ans:* 31 balloons.

10. A dry cleaner received 68 items of clothing and cleaned 45 of them the same day. How many items are left to be cleaned? *Ans:* 23 items.

11. The dry cleaner received 58 blouses and cleaned 36 the same day. How many more are left to be cleaned? *Ans:* 22 blouses.

12. Out of 39 lamp posts on our street, only 26 are lit. How many are not? *Ans:* 13 lamp posts.

13. A construction set has 80 parts.
    a) How many parts will be left if it takes 36 parts to make a car? *Ans:* 44 parts.
    b) How many parts will be left if it takes 45 parts to make a truck? *Ans:* 35 parts.
    c) How many parts will be left if it takes 56 parts to make a crane? *Ans:* 24 parts.

14. A farmer bought 50 chickens. If 34 were hens, how many roosters did he buy? *Ans:* 16 roosters.

15. The farmer also bought a goat and pig for $90. If the pig cost $55, how much did the goat cost? *Ans:* $35.

16. Altogether 44 pirates climbed a ship. If 35 pirates were guys, how many girl-pirates were on the ship? *Ans:* 9 girl-pirates.

17. A desk made of metal and wood weighs 55 pounds. If there are 16 pounds of metal in the desk, how much wood was used to make the desk? *Ans:* 39 pounds.

18. Two pirates together hid 64 gold and silver coins in a chest.
    a) If one pirate put 35 coins, what was the other's share? *Ans:* 29 coins.
    b) If 26 coins were gold, how many silver coins did they hide? *Ans:* 38 coins.

19. Missy can sew 25 dresses in one month, Liz can sew 26. How many dresses can they sew together? *Ans:* 51 dresses.

20. Before Mona got sick, she weighed 84 pounds. Now she weighs 76 pounds. How much weight did she lose? *Ans:* 8 pounds.

21. The boat with the motor cost $62. The boat cost $35. What's the price of the motor? *Ans:* $27. Motors for boats often cost as much as the boat itself.

22. Mr. Hasty bought a watch for $62 and quickly sold it for $56. How much money did he lose? *Ans:* $6.

23. In a building, 66 apartments are occupied by tenants. How many are still empty if there are 83 apartments in the building? *Ans:* 17 apartments.

24. Out of 83 apartments, 55 are one-bedrooms and the rest are two-bedrooms. How many two bedroom apartments are in the building? *Ans:* 28 apartments.

25. Yesterday, a rose bush in front of the window had 34 roses. Today there are only 16 roses. How many fell off? *Ans:* 18 roses.

26. In the morning there were 64 cars on the parking lot. By noon there were 26. How many cars left? *Ans:* 38 cars.

27. In a nursery, a nurse prepared 31 bottles of milk and gave 16 bottles to the babies for breakfast. How many bottles were left for lunch? *Ans:* 15 bottles.

28. In a nursery, a nurse prepared 31 diapers and used 15 in the morning. How many diapers were left for the afternoon? *Ans:* 16 diapers.

29. In a animal shelter with 42 puppies, 15 puppies were crying in the morning. How many were quiet? *Ans:* 27 puppies.

30. In a animal shelter with 42 puppies, 26 puppies fell asleep after eating their lunch. How many were awake? *Ans:* 16 puppies.

31. In a 51-camel caravan, 25 camels carry goods and the rest carry riders. How many camels have riders? *Ans:* 26 camels.

32. Ms. Harriet Fingers can type 61 words per minute and Mr. Al Thumbs can type 15 words per minute. How much faster is Ms. Fingers than Mr. Thumbs? *Ans:* 46 words per minute.

33. Ms. Fingers has to type 71 pages. She divided her job into two parts. If the first part has 36 pages, how big is the second part? *Ans:* 35 pages.

34. A company with 55 employees moved 36 people to a new location. How many employees stayed behind? *Ans:* 19 employees.

35. Out of 36 people who moved, 25 bought a house and the rest rented. How many people rented? *Ans:* 11 people.

36. A 64-seat theater sold 55 tickets for the show. How many seats were empty? *Ans:* 9 seats.

37. Out of 55 spectators, only 46 stayed until the end of the show. How many left before the show was over? *Ans:* 9 people.

38. Marina swam 94 laps in the pool, 35 with breast strokes and the rest with back strokes. How many laps did she swim with back strokes? *Ans:* 59 laps.

39. Although Noah planned to do 42 math problems, he was able to do only 26. How many math problems were too hard for him? *Ans:* 16 problems.

40. A redwood tree is 83 feet tall; a cedar is 36 feet shorter. How tall is the cedar? *Ans:* 47 feet.

41. Lola has three quarters in her pocket, or 75 cents.
a) How much would it be if she spends a quarter (25 cents)?
Ans: 50 cents.
b) How much will be left if she spends 50 cents? *Ans:* 25 cents.

42. Mr. Jay planted 100 trees but only 86 survived under the drought. How many didn't make it? *Ans:* 14 trees. Drought is period during which the expected amount of rain does not fall and the ground becomes dry and thirsty.

43. A hotel purchases 100 bars new soap bars every week but uses only 25.
    a) How many bars of soap were left after the first week?
    Ans: 75 bars.
    b) How many bars of soap remained after the second week?
    Ans: 50 bars.
    c) How many bars of soap were there after the third week?
    Ans: 25 bars.

44. A thief stole 90 watches but dropped 66 of them while running away from the police. How many did he escape with?
    Ans: 24 watches.

45. In the hospital, 65 nurses were divided in two shifts. If the day shift has 36 nurses, how many are in the night shift?
    Ans: 29 nurses.

46. To get to work, Sonia walks 25 minutes, and than takes a bus ride that is 6 minutes longer than her walk. How long does it take her to get to work? *Ans:* 56 minutes.
    Solution: It is best to see it as a 25 minute walk and then another 25 minute on the bus, which is 50 minutes and then 6 more minutes for a total of 56 minutes.

47. The first grade has 36 students and the second grade has 15 more than the first. How many students are in both grades?
    Ans: 87 students.
    Solution: Again, it is 36 plus 36 which is 72 and then 15 more or 87 students.

48. Saturday, Nina read 45 pages. Sunday she read 26 less. How many pages did she read in two days? *Ans:* 64 pages.
    Solution: Double of 45 is 90. Then subtract 26 from that.

49. In May, a store sold 43 bicycles, in June they sold 16 less. How many bicycles did they sell in the two months? *Ans:* 70 bicycles

50. A tailor had two pieces of fabric: one was 61 feet long and the other 46 feet less than that. How many feet of fabric were in two pieces together? *Ans:* 76 feet (61 - 46 = 15, 61 + 15 = 76).

ADDING DOUBLE-DIGIT NUMBERS ENDING IN 7 OR 8

EXERCISE I

18 + 15 = 33	48 + 11 = 59	35 - 16 = 19	98 - 48 = 50
28 + 16 = 44	18 + 15 = 33	67 - 17 = 50	75 - 38 = 37
27 + 25 = 52	38 + 25 = 63	68 - 28 = 40	35 - 28 = 7
38 + 35 = 73	28 + 50 = 78	67 - 27 = 40	55 - 18 = 37
28 + 45 = 73	48 + 13 = 61	53 - 26 = 27	45 - 18 = 27
38 + 55 = 93	38 + 45 = 83	62 - 38 = 24	55 - 28 = 27
48 + 17 = 65	38 + 30 = 68	62 - 48 = 14	65 - 38 = 27
58 + 16 = 74	45 - 18 = 27	76 - 47 = 29	45 - 18 = 27

EXERCISE II

- To which number do you add 7 to make 18? *Ans:* 11
- To which number do you add 8 to make 30? *Ans:* 22
- To which number do you add 8 to make 37? *Ans:* 29
- To which number do you add 17 to make 46? *Ans:* 29
- To which number do you add 18 to make 88? *Ans:* 70

- To which number do you add 17 to make 55? *Ans:* 38
  - To which number do you add 8 to make 64? *Ans:* 56
  - To which number do you add 8 to make 75? *Ans:* 67
  - To which number do you add 7 to make 54? *Ans:* 47
  - To which number do you add 17 to make 33? *Ans:* 16
  - To which number do you add 17 to make 62? *Ans:* 45
  - To which number do you add 18 to make 41? *Ans:* 23
  - To which number do you add 18 to make 81? *Ans:* 63
  - To which number do you add 17 to make 82? *Ans:* 65

WORD PROBLEMS

1. In a basketball game we scored 18 points less than the guest team. How many points did we score, if the guests scored 90 points? *Ans:* 72 points.

2. It took Sarah 17 minutes to walk to the store and 8 minutes longer to walk back. How long did it take for the whole trip? *Ans:* 42 minutes.
   Solution: Walking back from the store took 17 (minutes) + 8 (minutes) = 25 (minutes). The whole trip took 17 (minutes) + 25 (minutes) = 42 (minutes). *Another way:* Double of 17 is 34 and then add 8 to that to get 42 minutes.

3. Cinderella washed 17 dresses and 16 skirts. How many of both, dresses and skirts, did she wash? *Ans:* 33 dresses and skirts.

4. A prince tried the glass slipper on 18 girls in one college and 15 girls in another collage. On how many girls tried on the slipper? *Ans:* 33 girls. Solution: We can double the smaller number which is 15. We get 30. Now we can add to it the difference between the larger and smaller number, which is 3. So the answer is 33.

5. In the book, stories takes 27 pages and fables take 28. How many pages are in the book? *Ans:* 55 pages.

6. Pinocchio told 34 big lies and 28 small ones. How many lies did Pinocchio tell? *Ans:* 62 lies.

7. It took 18 minutes to find a ladder and another 18 minutes to bring the kitten down from the tree. How long did the whole rescue last? *Ans:* 36 minutes.

8. In the park, 17 children were playing "Simon Says." Then 7 more girls and 8 more boys joined the game. How many children are playing now? *Ans:* 32 children (17 + 7 = 24, 24 + 8 = 32).

9. Sandy delivered 27 newspapers a day. How many newspapers does he deliver in 2 days? *Ans:* 54 newspapers.

10. Last year 26 new houses were built in our town. This year they built 37 more houses. How many houses were built in two years? *Ans:* 63 houses.

11. Gilmore has 26 pairs of shoes with shoelaces and 17 pairs without. How many pairs of shoes does he have altogether? *Ans:* 43 pairs.

12. There are 48 green apples in one bag and 47 red ones in the other. How many apples are in both bags? *Ans:* 95 apples.

13. There are 38 paperback books on one table and 37 hardback books on the other. How many books are on both tables? *Ans:* 75 books.

14. A team had 25 hockey sticks and bought 17 more. How many old and new sticks does the team have? *Ans:* 42 hockey sticks.

15. The first show lasted 28 minutes and the second was 33 minutes long. How long did both shows last? *Ans:* 61 minutes or 1 hour and 1 minute.

16. From a vegetable patch, Julio picked 33 long cucumbers and 27 short ones. How many cucumbers of all sizes did Julio pick? *Ans:* 60 cucumbers.

17. A busy pediatrician saw 27 patients in the morning and 35 in the afternoon. How many patients did he see that day? *Ans:* 62 patients.

18. One trash can holds 34 gallons, the other 16 gallons more than the first. How much trash can both cans hold? *Ans:* 84 gallons. Solution: The second trash can holds 34 (gallons) + 16 (gallons) = 50 (gallons) of trash. Together, both cans hold 34 (gallons) + 50 (gallons) = 84 (gallons).

19. One trash can costs $27, another costs $37. How much do both cost? *Ans:* $64.

20. At night in a desert, 17 giraffes and 33 zebras came to a waterhole. How many animals came for a drink? *Ans:* 50 animals.

21. Little Red Riding Hood has in her basket 18 muffins and 12 biscuits. How many muffins and biscuits did she carry? *Ans:* 30 muffins and biscuits.

22. A squirrel preparing for winter stored 29 seeds and 28 nuts. How many seeds and nuts did the squirrel store? *Ans:* 57 seeds and nuts.

23. Yesterday the children solved 37 problems. Today, they did 8 more than yesterday. How many problems did they solve in 2 days? *Ans:* 82 problems.
    Solution: Today children solved 37 (problems) + 8 (problems) = 45 (problems). Both days children solved 37 (problems) + 45 (problems) = 82 (problems).
    Another way: Double of 37 is 74. Add 8 to that to get 82.

24. If Inga was 14 years old 17 years ago, how old is she now? *Ans:* 31 years old.

25. If Mr. Bill was 49 years old 17 years ago, how old is he now? *Ans:* 66 years old.

26. If 68 years ago my grandpa was 13, how old is he now? *Ans:* 81 years old.

27. A small village by the sea had 38 inhabitants. In the summer, 54 visitors came to the village. How many people were in the village during the summer? *Ans:* 92 people.

28. Vaclav hung a bird feeder outside his window and counted 32 finches and 38 sparrows. How many birds came to his feeder? *Ans:* 70 birds.

29. A brother and his sister were working in the garden. The brother pulled 46 weeds while the sister pulled 48. How many weeds did they pull together? *Ans:* 94 weeds.

30. The brother and the sister planted lemon trees in their orchard. The brother planted 27 trees and the sister planted 26. How many trees did they plant together? *Ans:* 53 trees.

31. At the convention center, one salesman passed 37 business cards and the other passed 34 cards. How many business cards did they both pass out? *Ans:* 71 cards.

32. Using a nutcracker, Morris cracked 25 walnuts and 67 hazelnuts. How many nuts did Morris crack? *Ans:* 92 nuts.

33. Playing Scrabble, Lin made 19 words and May came up with 17. How many words did they both make? *Ans:* 36 words.

34. Playing Scrabble, Lin earned 47 points and May earned 48. How many points did they both earn? *Ans:* 95 points.

35. Samantha, looking at the sky, counted 17 clouds shaped like a sausage and 23 in the shape of a muffin. How many clouds of both shapes did she count? *Ans:* 40 clouds. Do you think she might have been hungry?

36. If Madison has 22 Lego pieces, Georgia has 17 pieces, and Eva has 18, how many pieces are there in all? *Ans:* 57 pieces.

37. The crowd at the baseball game yelled "Hurray!" 35 times and "Boo!" 38 times. How many times did the crowd yell? *Ans:* 73 times.

38. Colin made 28 errors while rehearsing his lines for a play. If he makes 17 more errors, his acting career will be over. How many errors is he allowed to make? *Ans:* 45 errors.

39. Colin had to memorize 87 lines of dialog for the play. He remembered most of them except 13. How many lines did he know well? *Ans:* 74 lines.

40. A magazine editor checked 29 articles and has to check 17 more. How many articles does he have to check? *Ans:* 46 articles.

41. The newspaper will have 44 articles. The editor had decided on 27 of them but is not sure of the rest. How many is he not sure about? *Ans:* 17 articles.

42. A group of 38 anteaters saw a line of 29 ants. Everyone gets only one ant; how many did not get an ant? *Ans:* 9 anteaters.

43. An anteater saw a line of 56 ants. He gobbled up 49 of them and then a bird swooped down and scooped up the rest. How many ants did the bird get? *Ans:* 6 ants.

44. One truck carried 18 foosball tables; the other had 17 more than the first. How many foosball tables did both trucks carry? *Ans:* 53 tables.
    Solution: The second truck had 18 (tables) + 17 (tables) = 35 (tables). Both trucks had 18 (tables) + 35(tables) = 53 (tables).

45. Gigi has $17, Heather has $19, and Irene has $13. How much money do all three girls have? *Ans:* $49.

46. Solution: 17 + 19 + 13 is easier to group by first adding 17 and 13, and then 19. So 17 + 13 = 30, and 30 + 19 = 49. Much easier!

47. There are 27 fish in the first fish tank and 29 more fish in the second. How many fish are in both fish tanks? *Ans:* 83 fish.
    Solution: In the second fish tank, there are 27 (fish) + 29 (fish) = 56 (fish). In both tanks, there are 27 (fish) + 56 (fish) = 83 (fish).

48. The principal asked all 7-year old kids to stand to the right, all 8-year old kids to stand in the middle and all 9-year old kids to stand on the left. There were 17 7-year olds; 18 8-year olds and 19 9-year olds. How many kids were at the assembly?
    Ans: 54 kids.

49. The band teacher told everyone that they needed at least 37 kids to make the band. Only 18 kids had signed up so far. How many more kids need to sign up to complete the band?
    Ans: 19 kids more.

50. The ship going down the coast dropped off 17 tons of cargo at port A, then it dropped off 15 tons at port B. If it had 50 tons of cargo to start with, how much cargo does it still have on board? *Ans:* 18 tons.

51. The ship had 37 crew members when it started. Seven of them got off at port A and 8 crew members got off at port B. How many crew are still on board? *Ans:* 22 crew members.

52. There were 17 white roses on the table, 16 pink roses and 7 red roses. How many roses were there on the table? *Ans:* 40 roses.

13

SUBTRACTING DOUBLE-DIGIT NUMBERS ENDING IN 7 OR 8

EXERCISE I

- Count forward from 0 to 100 and then backward to 0 by 5.
  (0, 5, 10, 15, etc., 100, 95, 90, 85, etc., 0)

- Count forward from 0 to 96 and then backward to 0 by 6.
  (0, 6, 12, 18, etc., 96, 90, 84, 78, etc., 0)

- Count forward from 0 to 77 and then backward to 0 by 7.
  (0, 7, 14, 21, etc.)

- Count forward from 0 to 80 and then backward to 0 by 8.
  (0, 8, 16, 24, etc.)

EXERCISE II

- To which number do you add 7 to make 29? *Ans:* 22

- To which number do you add 7 to make 30? *Ans:* 23

- To which number do you add 7 to make 34? *Ans:* 27

- To which number do you add 7 to make 46? *Ans:* 39

- To which number do you add 7 to make 85? *Ans:* 78

- To which number do you add 7 to make 53? *Ans:* 46
- To which number do you add 7 to make 41? *Ans:* 34
- To which number do you add 7 to make 72? *Ans:* 65
- To which number do you add 8 to make 54? *Ans:* 46
- To which number do you add 8 to make 33? *Ans:* 25
- To which number do you add 8 to make 67? *Ans:* 59
- To which number do you add 8 to make 41? *Ans:* 33
- To which number do you add 8 to make 55? *Ans:* 47

EXERCISE III

18 - 15 = 3	55 - 38 = 17	38 - 25 = 13	45 - 18 = 27
28 - 16 = 12	48 - 17 = 31	50 - 28 = 22	35 - 16 = 19
27 - 25 = 2	58 - 16 = 42	48 - 13 = 35	67 - 17 = 50
38 - 35 = 3	48 - 11 = 37	45 - 38 = 7	68 - 28 = 40
45 - 18 = 27	18 - 15 = 3	38 - 30 = 8	28 - 18 = 10

WORD PROBLEMS

1. In a basketball game we scored 18 points less than the guest team. How many points did we score if the guests scored 90 points? *Ans:* 72 points.

2. Out of 100 recruits, 48 went to the boot camp right away. How many had to wait for their turn? *Ans:* 52 recruits.

3. The second grade students planted 31 trees. The first graders planted 7 fewer. How many trees did the first graders plant? *Ans:* 24 trees.

4. One bus carried 50 tourists, the other had 17 less. How many tourists were on the two buses? *Ans:* 83 tourists.
   Solution: The second bus had 50 (tourists) - 17 (tourists) = 33 (tourists). Next, two buses have 50 (tourists) + 33 (tourists) = 83 (tourists).

5. To go over the mountain takes 44 hours. The path around the mountain is 17 hours longer. How long does it take to go around the mountain? *Ans:* 61 hours.

6. There were 38 passengers on the train and at the next station 26 more boarded the train. How many passengers were on the train then? *Ans:* 64 passengers.
   a) At the next station, 17 passengers got off. How many stayed on? *Ans:* 47 passengers.
   b) Then 28 passengers got off the train. How many kept going? *Ans:* 19 passengers.

7. In the morning on Mother's Day, a flower shop had 70 roses for sale. By noon, they had sold all but 17 roses. How many roses did they sell in the morning? *Ans:* 53 roses.

8. If in 18 years my teacher will be 50, how old is she now? *Ans:* 32 years old.

9. If in 38 years Inez will be 44, how old is she now? *Ans:* 6 years old.

10. If in 77 years my sister will be 81, how old is she now? *Ans:* 4 years old.
    How old will you be in 9 years?

11. A suit (jacket and pants) costs $83. The jacket alone costs $57. How much did the pants cost? *Ans:* $26.

12. It took 56 boards to build a play house; 18 boards for the roof and the rest for the walls. How many boards were used for the walls? *Ans:* 38 boards.

13. Frank and George are 54 miles apart. They are driving toward each other. If Frank drove 28 miles, how far did George drive when they met? *Ans:* 26 miles.

14. There were 75 passengers and 18 crew members on board a small cruise boat. How many people were on board? *Ans:* 93 people.

15. There were 75 passengers on board. At Ensenada, 67 passengers went ashore. How many stayed on board? *Ans:* 8 passengers. Ensenada is a port town on the Western side of Mexico.

16. Of the 75 passengers on board, at dinner time 48 passengers ordered fish. How many didn't order the fish? *Ans:* 27 passengers.

17. Mr. Whiff gave his wife 64 ounces of perfume. After a month, there were only 17 ounces left in the bottle. How many ounces of perfume did Mrs. Whiff use? *Ans:* 47 ounces.

18. A two-story school bought 73 light bulbs and used 38 of them on the 1st floor. How many light bulbs are to be used on the second floor? *Ans:* 35 light bulbs.

19. Chuck bought a bag with 52 hot chili peppers and quickly ate 18 of them. How many chili peppers were still in the bag when he screamed? *Ans:* 34 chili peppers and they are as hot as the first 18.

20. Out of 84 blocks from the box, little Timmy took 48 to build a tower. How many blocks did he not use? *Ans:* 36 blocks.

21. While on bird-watching tour, Latisha counted 43 birds on a lake. She recognized only 27 of them. How many birds on the lake were unfamiliar to her? *Ans:* 16 birds.

22. Carlos has $47 in his pocket and $28 in the wallet. How much money does he have? *Ans:* $75.

23. Two bird feeders in our backyard attracted 56 birds. If the larger feeder brought in 28 birds, how many birds came to the smaller one? *Ans:* 28 birds. Can birds tell the difference between a big and a small bird feeder?

24. At a egg farm, 47 chickens hatched out of 70 eggs. How many eggs didn't hatch? *Ans:* 23 eggs.

25. Lucy borrowed $100 and returned $58. How much money does she still owe? *Ans:* $42.

26. If a skateboarder got a total of 50 stitches on both of his scratched up legs. If he got 27 stitches on the right leg, how many stitches were needed for his left leg? *Ans:* 23 stitches.

27. A secret password has 50 characters made of letters and digits. If 18 characters are digits, how many letters are in the password? *Ans:* 32 letters.

28. For 80-cent postage, Trevor bought 2 stamps. If one stamp is 57 cents, how much is the other stamp? *Ans:* 23 cents.

29. A necklace has 61 pearls; 27 pearls are black and the rest are white. How many white pearls does the necklace have?
Ans: 34 pearls.

30. A group of 55 students went to a Mystery Museum but only 47 came out. How many students got lost in the museum?
Ans: 8 students. Don't worry they were found the next day.

31. A mom and her daughter together are 63 years old. If mom is 47 years old, how old is the daughter? **Ans:** 16 years old.

32. Robin Hood had 73 golden coins and gave away 67. How many did he keep for himself? **Ans:** 6 coins.

33. A fisherman caught 41 fish. He sold 18 fish right away and froze the rest. How many fish did he freeze? **Ans:** 23 fish.

34. There were 55 planes still at the tarmac after 17 planes took off. How many planes were there before the 17 planes departed?
Ans: 72 planes. Tarmac is the big cemented area where airplanes park at the airport.

35. There were 83 passengers on the plane when it arrived. At the airport 28 passengers got off. How many passengers stayed on the plane? **Ans:** 55 passengers.

36. A golf team brought 64 balls to the driving range. By the end of the day they had only 47 balls. How many balls did the team lose? **Ans:** 17 balls.

37. In the morning, 35 fishing boats left the harbor. Then 27 more boats left. How many fishing boats left the harbor?
Ans: 62 boats.

38. There were 63 yachts in the harbor, then 38 yachts left. How many yachts are moored (stayed fastened) in the harbor?
Ans: 25 yachts.

39. Out of 63 yachts in the harbor, 27 are big yachts and the rest are small. How many small yachts are in the harbor?
Ans: 36 yachts.

40. Eighty-three sailors stayed overnight at the Harbor Hotel. Of these 47 sailors had beards and the rest didn't. How many sailors were beardless? **Ans:** 36 sailors.

41. In the restaurant, 28 sailors out of the 83, ordered fish, but the others ate veggies. How many sailors ate veggies? *Ans:* 55 sailors. On a long trips at sea, in the past sailors used to get sick because they were not able to eat fruits and vegetables.

42. A company with 63 employees laid off 27 people. How many people remain? *Ans:* 36 people. What does a layoff mean?

43. There are 29 lamp posts on one street and 27 on the other. Of these 18, lamp posts are not working. How many do work? *Ans:* 38 lamp posts.
Solution: On both streets, there are 29 (lamp posts) + 27 (lamp posts) = 56 (lamp posts).Working lamp posts are 56 (lamp posts on both streets) - 18 (lamp posts that don't work) = 38 (lamp posts).

44. Nora bought two books, one was $18 and the other was $28. She paid for them with a $100 bill. How much in change did she receive? *Ans:* $54.
Solution: There are two ways to solve this problem. First, we can add the price of two books together, $18 + $28 = $46. Next, we calculate the change $100 - $46 = $54.
Second way: we can subtract one book at a time from $100. Thus, $100 - $18 = $82, then $82 - $28 = $54.

45. Phil bought two stamps, costing 40 cents and 38 cents, and paid with a $1 bill (100 cents). How much change did he get? *Ans:* 22 cents.

46. From the box with 100 pencils the teacher took out 28 pencils for the first period and 33 for the second. How many pencils were left in the box? *Ans:* 39 pencils.

47. There were 83 guests at a Birthday party. If 47 guests are local, how many guests came from out of town? *Ans:* 36 guests.

48. At the party, all 83 guests ordered drinks; 57 guests ordered iced tea and the rest had lemonade. How many guests drank lemonade? *Ans:* 26 guests.

49. At the party Gordon ate 62 corn chips. He dipped only 27 chips in salsa. How many chips did he eat without dipping? *Ans:* 35 chips.

50. The bride and the groom received 42 gifts; there were 17 salad bowls and the rest were flower vases. How many vases did they get? *Ans:* 25 vases.

ADDING TWO DOUBLE-DIGIT NUMBERS UP TO 100

EXERCISE I

- Count from 1 to 50 skipping 2 numbers; or, in other words, by adding 3 (1, 4, 7, 10, 13, 16, etc.)

- Count from 0 to 50 skipping 2 numbers; or, in other words, by adding 3 (0, 3, 6, 9, 12, 15, etc.)

- Count from 2 to 50 skipping 2 numbers; or, in other words, by adding 3 (2, 5, 8, 11, 14, 17, etc.)

HOW TO SOLVE

First Problem: 36 + 23 =?

First Solution: Step 1: 23 = 20 + 3
Step 2: 36 + 20 = 56, and then 56 + 3 = 59

The answer: 36 + 23 = 59.

Second Solution: Step 1: 36 = 30 + 6 and 23 = 20 + 3
Step 2: Let's add tens first, 30 + 20 = 50.
Now we'll add ones: 6 + 3 = 9.
Step 3: 50 + 9 = 59

The answer: 36 + 23 = 59.

EXERCISE II

56 + 15 = 71	56 + 43 = 99	80 + 16 = 96	63 + 13 = 76
37 + 16 = 53	46 + 14 = 60	45 + 27 = 72	32 + 67 = 99
45 + 18 = 63	43 + 53 = 96	35 + 34 = 69	14 + 25 = 39
67 + 25 = 92	32 + 15 = 47	32 + 24 = 56	19 + 11 = 30
33 + 11 = 44	71 + 16 = 87	38 + 33 = 71	28 + 8 = 36
57 + 23 = 80	47 + 44 = 91	44 + 15 = 59	39 + 18 = 57
82 + 16 = 98	38 + 21 = 59	62 + 20 = 82	49 + 18 = 67
45 + 17 = 62	59 + 29 = 88	68 + 14 = 82	68 + 32 = 100

EXERCISE III

56 - 15 = 41	46 - 14 = 32	35 - 34 = 1	19 - 11 = 8
37 - 16 = 21	57 - 53 = 4	32 - 24 = 8	28 - 8 = 20
45 - 18 = 27	32 - 15 = 17	38 - 33 = 5	39 - 18 = 21
67 - 25 = 42	71 - 16 = 55	44 - 15 = 29	49 - 18 = 31
33 - 11 = 22	47 - 44 = 3	62 - 20 = 42	68 - 32 = 36
57 - 23 = 34	38 - 21 = 17	68 - 14 = 54	
82 - 16 = 66	59 - 29 = 30	63 - 13 = 50	
45 - 17 = 28	80 - 26 = 54	52 - 27 = 25	
56 - 43 = 13	45 - 27 = 18	32 - 25 = 7	

EXERCISE IV

- Which 2 numbers added together would make 26?
  Give at least 3 examples. (12 + 14, 19 + 7, 8 + 18).

- Which 2 numbers added together would make 42?
  Give at least 3 examples.

- Which 2 double-digit numbers added together would make 50?
  Give at least 3 examples.

- Which 3 double-digit numbers added together would make 42?
  Give at least 3 examples.

- Which 2 double-digit numbers added together would make 42?
  Give at least 3 examples.

- Which 3 double-digit numbers added together would make 45?
  Give at least 3 examples.

- Which 3 double-digit numbers added together would make 50?
- Which 4 double-digit numbers added together would make 50?

WORD PROBLEMS

1. A man wants to buy a jacket which costs $85. He has only $53. How much more money does he need to be able to buy the jacket? *Ans:* $32.

2. Nina weighed 55 lb and gained 13 pounds during the summer. What's her weight now? *Ans:* 68 lb.

3. It takes 27 minutes to walk to the bus stop and then a 21-minute ride on the bus. How long is the trip? *Ans:* 48 minutes.

4. Two boys each brought 36 marbles. How many marbles did they both bring? *Ans:* 72 marbles.

5. One class has 26 students, the other has 29. How many students are in both classes? *Ans:* 55 students.

6. An airline company has 36 passenger planes and 14 cargo planes. How many planes does the airline have? *Ans:* 50 planes.

7. There were 46 home team fans and 35 guest team fans at the game. How many fans were at the game? *Ans:* 81 fans.

8. On a narrow path, 33 red ants met 27 black ants. How many ants met? *Ans:* 60 ants.

9. A mailman delivered 45 letters and 37 parcels. How many pieces of mail did he deliver? *Ans:* 82 pieces.

10. Every day, 26 passenger and 15 cargo trains pass by a small station. How many trains pass by the station every day? *Ans:* 41 trains.

11. To an ice box with 27 pounds of frozen meat, Bianca added 19 pounds of fish. How much meat is in the box now? *Ans:* 46 pounds.

12. Running away from a cat, a mouse scampered 69 feet and then another 14 feet. How far did the mouse run? *Ans:* 83 feet. To scamper means to run away quickly.

13. One swallow caught 49 flies, another caught 29. How many flies did they catch together? *Ans:* 78 flies.

14. A senator made 16 short and 57 long speeches. How many speeches did she make? *Ans:* 73 speeches.

15. A farm has 48 sheep and 16 goats. How many animals does the farm have? *Ans:* 64 animals.

16. A store sold 35 bicycles and 44 scooters. How many of both did it sell? *Ans:* 79 bicycles and scooters.

17. Giant's mom made him a salad with 54 tomatoes and 38 cucumbers. How many of both vegetables did she use? *Ans:* 93 tomatoes and cucumbers. His mom wants him to eat a lot of vegetables.

18. A tailor shop used 72 buttons and 16 zippers. How many buttons and zippers did it use? *Ans:* 88 buttons and zippers.

19. A dragon had 3 heads. The first head threw 27 fireballs, the second threw 18, and the third was asleep. How many fire balls all three heads threw? *Ans:* 45 fireballs.

20. It took Drew 28 minutes to write a report and another 13 minutes to correct the errors. How many minutes did it take to finish the report? *Ans:* 41 minutes.

21. A shop printed 48 posters for a game, ran out of all of them and then had to print 42 more. How many posters were printed? *Ans:* 90 posters.

22. At a farm, 23 cabbages were planted in the first row and 39 in the second. How many cabbages were planted in both rows? *Ans:* 62 cabbages.

23. If one cargo train had 37 cars and the other train had 36 cars, how many cars did they both have? *Ans:* 73 cars.

24. My uncle Morris is 48 years old.
    a) How old is he going to be in 12 years? *Ans:* 60 years old.
    b) How old is he going to be in 16 years? *Ans:* 64 years old.
    c) How old is he going to be in 19 years? *Ans:* 67 years old.
    d) How old is he going to be in 25 years? *Ans:* 73 years old.

25. My aunt Jamie is 55 years old.
    a) How old is she going to be in 14 years? *Ans:* 69 years old.
    b) How old is she going to be in 17 years? *Ans:* 72 years old.
    c) How old is she going to be in 19 years? *Ans:* 74 years old.
    d) How old is she going to be in 24 years? *Ans:* 79 years old.

26. A secret agent had 26 grey raincoats and 28 brown ones. How many raincoats did he have? *Ans:* 54 raincoats.

27. Each secret agent's raincoat had 17 regular pockets and 27 secret ones. How many pockets did their raincoats have? *Ans:* 44 pockets.

28. A secret agent had 29 regular pens and 24 pens with invisible ink. How many pens did the agent have? *Ans:* 53 pens.

29. A secret agent had 34 fake beards and 16 wigs. How many wigs and beards did he have? *Ans:* 50 wigs and beards.

30. Omar paid $37 for a prize hamster and $19 for a cage. How much did he pay for both? *Ans:* $56.

31. A Magic Carpet store sold 48 magic carpets and 44 ordinary doormats. What is the total number of carpets and doormats the store sold? *Ans:* 92 carpets and mats.

32. One bottle has 64 ounces of soda and the other bottle has 32 ounces. How much soda is in both bottles? *Ans:* 96 ounces.

33. There are 13 miles between towns A and B, and 29 miles between B and C. What's the distance between the towns A and C? *Ans:* 42 miles.

34. Ken will use 18 cans of peaches and 25 cans of pears in his gigantic cake. How many total fruit cans will Ken use? *Ans:* 43 cans.

35. Two brothers left from home in opposite directions. One drove 26 miles, the other 28 miles. How far were they from each other? *Ans:* 54 miles.

36. The temperature outside is 46 degrees.
    a) If the temperature moves up 13 degrees, what will it be? *Ans:* 59 degrees.

b) If the temperature moves up 17 degrees, what will it be?
Ans: 63 degrees.
c) If the temperature moves up 19 degrees, what will it be?
Ans: 65 degrees.
d) If the temperature moves up 26 degrees, what will it be?
Ans: 72 degrees.

37. The guitar costs $79 and the strings $12. How much is the guitar with the strings attached? *Ans:* $91.

38. Two men walked from the same spot in opposite directions. The first man walked 36 miles; the second man walked 25 miles. How many miles are between them now? *Ans:* 61 miles.

39. A school class of 38 boys and 37 girls went to see a play. How many children went to see the play? *Ans:* 75 children.

40. There are 75 children in the theater and 15 seats are empty. How many seats are in the theater? *Ans:* 90 seats.

41. To a wire 25 feet long, a 28-foot long piece was spliced. What's the length of both wires? *Ans:* 53 feet. To splice means to join.

42. There are 34 daisies in one bunch and 8 daisies in the other. How many daisies are in both bunches? *Ans:* 42 flowers.

43. Mimi holds 68 cents in one hand and 16 cents in the other. How much money is in both her hands? *Ans:* 84 cents

44. The first chapter of a book ended on page 46, the second chapter ended 25 pages further. On what page does the second chapter end? *Ans:* page 71.

45. Xenia picked 57 apples from a tree and 17 apples from the ground. How many apples did she pick? *Ans:* 74 apples.

46. There are 43 pins and 38 needles in a pincushion. How many pins and needles are there? *Ans:* 81 pins and needles.

47. Thirty-four cobras and 38 crocodiles lived near a swamp. How many cobras and alligators lived there? *Ans:* 72 creatures.

48. A chef sliced a cucumber into 27 very thin slices and another into 15 fat slices. How many slices did he cut? *Ans:* 42 slices.

49. Cameron cut a pipe in two parts of 47 and 34 inches long. How long was the pipe before being cut? *Ans:* 81 inches.

50. There are 28 steps from my home to the sidewalk and 33 steps to the crosswalk. How many steps is my home to the crosswalk? *Ans:* 61 steps.

51. A hungry giant found a pizza delivery truck with 26 cheese and 35 pepperoni pizzas and ate them all. How many pizzas did the giant eat? *Ans:* 61 pizzas.

52. There were 28 gallons of water in the tank and Troy added 25 gallons to fill it up. How many gallons can the tank hold? *Ans:* 53 gallons.

53. A man fed 39 ducks and has 36 more ducks to feed. How many ducks does he have to feed? *Ans:* 75 ducks.

15

DOUBLING AND TRIPLING DOUBLE-DIGIT NUMBERS

Doubling means adding two of the same numbers together. Tripling means adding three of the same numbers together.

▶ *A simple rule: When doubling (or tripling) double digit numbers it might be easier to add the tens first, then add the ones and then add the sums together.*

32 + 32 = ?   30 + 30 + 2 + 2 = 64
27 + 27 = ?   20 + 20 + 7 + 7 = 54

EXERCISE I

- Count from 1 to 50 skipping 3 numbers; or, in other words, by adding 4 (1, 5, 9, 13, 17, 21, 25, etc.)

- Count from 0 to 50 skipping 3 numbers; or, in other words, by adding 4 (0, 4, 8, 12, 16, 20, 24, etc.)

- Count from 2 to 50 skipping 3 numbers; or, in other words, by adding 4 (2, 6, 10, 14, 18, 22, 26, etc.)

- Count from 3 to 50 skipping 3 numbers; or, in other words, by adding 4 (3, 7, 11, 15, 19, 23, 27, etc.)

EXERCISE II

- Count backwards from 50 to 0 skipping every other number (50, 48, 46, 44, 42, 40, 38, etc.)
- Count backwards from 49 to 1 skipping every other number (49, 47, 45, 43, 41, 39, 37, etc.)

EXERCISE III

We use words like "double", "twice" "as much or twice as many", "two times", or "twofold" to tell that we need to add an amount to itself. Soon, we will be learning multiplication - a shortcut to addition - and will be saying "multiply by 2."

10 + 10 = 20	20 + 20 = 40	32 + 32 = 64	50 + 50 = 100
11 + 11 = 22	21 + 21 = 42	33 + 33 = 66	51 + 51 = 102
13 + 13 = 26	22 + 22 = 44	35 + 35 = 70	60 + 60 = 120
15 + 15 = 30	25 + 25 = 50	40 + 40 = 80	80 + 80 = 160
16 + 16 = 32	28 + 28 = 56	44 + 44 = 88	70 + 70 = 140
18 + 18 = 36	30 + 30 = 60	49 + 49 = 98	50 + 50 = 100

WORD PROBLEMS

1. If one tank holds 24 gallons of water, how much water would 2 tanks hold? *Ans:* 48 gallons.

2. If one juice can holds 16 ounces, how many ounces do two cans hold? *Ans:* 32 ounces.

3. One tent can hold 15 people.
   a) How many people can fit in 2 tents? *Ans:* 30 people.
   b) How many people can fit in 3 tents? *Ans:* 45 people.

4. If one squad has 11 soldiers, how many soldiers are in 3 squads? *Ans:* 33 soldiers. By the way, that makes a platoon.

5. If there are 33 soldiers in one platoon, how many are in 3 platoons? *Ans:* 99 soldiers.
   By the way, that makes a company. Several companies will make a battalion, and several battalions group to form a brigade.

6. Valerie has $33 and Vera has as much. How much money do they have together? *Ans:* $66.

7. Kelly rode his bicycle 24 miles to the next village and then back. How many miles did he travel? *Ans:* 48 miles.

8. How many wings do 41 birds have? *Ans:* 82 wings.

9. A company president sent out 32 emails and received a reply to all of them. How many emails were sent and received altogether? *Ans:* 64 messages.

10. Grandpa man kissed both cheeks of his 23 grandchildren. How many cheeks did he kiss? *Ans:* 46 cheeks.

11. How many ears do 21 rabbits have? *Ans:* 42 ears.

12. How many eyes do 34 raccoons have? *Ans:* 68 eyes.

13. How many shoes can 22 runners wear? *Ans:* 44 shoes.

14. How many gloves can 44 boxers wear? *Ans:* 88 gloves.

15. How many ends do 43 earthworms have? *Ans:* 86 ends.

16. How many feet do 31 flamingos have? *Ans:* 62 feet.

17. If one goat costs $42, how much do 2 goats cost? *Ans:* $84.

18. One soccer team has 11 players.
    a) How many players do 2 teams have? *Ans:* 22 players.
    b) How many players do 3 teams have? *Ans:* 33 players.

19. There are 22 players on the field during a soccer game.
    a) How many players are on the field during 2 games?
    Ans: 44 players.
    b) How many players are on the field during 3 games?
    Ans: 66 players.

20. There are 21 seats on a bus.
    a) How many seats are on 2 buses? *Ans:* 42 seats.
    b) How many seats are on 3 buses? *Ans:* 63 seats.

21. A zebra has 26 stripes on each side. How many stripes are on both sides? *Ans:* 52 stripes.

22. If one show lasts 31 minutes, how long will 2 shows last?
    Ans: 62 minutes.

23. If one show lasts 31 minutes, how long will 3 shows last?
    Ans: 93 minutes.

24. If there are 35 baseball caps in one box, how many are in 2 boxes? *Ans:* 70 caps.

25. If it takes 27 screws to build one cabinet, how many screws will it take for 2 cabinets? *Ans:* 54 screws.

26. The teacher asked Antonio to do 28 problems, but he did 28 more than asked. How many problems did he do?
Ans: 56 problems. Doing extra work is a good way to impress everyone.

27. There are 26 pairs of gloves in the storage. How many gloves are there? *Ans:* 52 gloves.

28. There are 26 pairs of boots in the storage. How many boots are in storage? *Ans:* 52 boots.

29. There are 36 pairs of gloves in the storage. How many gloves are there? *Ans:* 72 gloves.

30. There are 46 pairs of socks in the storage. How many socks are there? *Ans:* 92 socks.

31. There were 27 books on the shelf and a librarian put another 27. How many books are on the shelf now? *Ans:* 54 books.

32. The home team scored 28 points and the guests tied the game. How many points were scored? *Ans:* 56 points.

33. A bear slept for 29 days, woke up, and then slept 29 more days. How many days did the bear sleep? *Ans:* 58 days.

34. A cord was cut into 2 equal parts. Each part is 38 inches. How long was the cord before it was cut? *Ans:* 76 inches.

35. A squirrel split 25 nuts into halves. How many halves did it make? *Ans:* 50 halves.

36. A tailor received an order for 38 two-piece suits. How many pieces did he make? *Ans:* 76 pieces.

37. A boat rental shop has 45 boats. Each boat comes with 2 oars. How many oars are at the shop? *Ans:* 90 oars.

38. Jackson, a border collie, herded 37 cows and 37 sheep. How many farm animals did he herd? *Ans:* 74 animals.

39. It took 39 hours to get to the top of the mountain and 39 hours back. How long did the trip take? **Ans:** 78 hours.

40. One day lasts 24 hours.
    a) How many hours are in 2 days? **Ans:** 48 hours.
    b) How many hours are in 3 days? **Ans:** 72 hours.
    Solution: There are 2 ways to solve it. First way: 48 hours (in 2 days) + 24 hours (one more day) = 72 hours.
    Second Solution: We need to add three 24s together. 24 is made of 20 and 4. We add 20 + 20 + 20 = 60. Then, 4 + 4 + 4 = 12. Now, 60 + 12 = 72.

41. If 2 days have 48 hours, how many hours are in 4 days?
    Ans: 96 hours.
    Solution: If 2 days have 48 hours, then 4 days will last 48 + 48 (40 + 40 = 80; and 8 + 8 = 16; then 80 + 16 = 96).

42. If there are 25 sharks in the tank and each has 2 gills, how many gills do all the sharks have? **Ans:** 50 gills.

43. If one diver has 2 fins, how many fins do 27 divers have?
    Ans: 54 fins.

44. A rock star hired 26 porters to carry his bags to the tour bus. If each porter carried 2 bags, how many bags did they carry?
    Ans: 52 bags.

45. If each of 35 clocks has 2 hands (hours and minutes), how many hands do all the clocks have? **Ans:** 70 hands.

46. If each of 31 watches has 3 hands (hours, minutes, and seconds), how many hands are on all 31 watches?
    Ans: 93 hands.
    Solution: 31 watches have 31 hour hands, 31 minute hands, and 31 second hands; thus 31 + 31 + 31 = 93 (hands).

47. A security guard gets one day off for every day he works. If he worked 45 days, how many work and off days did he have altogether? **Ans:** 90 days.

48. There are just 2 players on each beach volleyball team. 49 teams showed to play in the tournament. How many players are on 49 teams? **Ans:** 98 players.

49. If one pair of pants has 3 pockets, how many pockets do 28 pants have? *Ans:* 84 pockets. (20 + 20 + 20 = 60; 8 + 8 + 8 = 24; 60 + 24 = 84).

50. If there are 27 runners in one competition, how many runners are in 3 competitions? *Ans:* 81 runners.

16

SUBTRACTING DOUBLE-DIGIT NUMBERS FROM NUMBERS UP TO 100

EXERCISE I

- Count down from 83 to 47 subtracting 6 (i.e., 83, 77, 71, 65, 59, 53, and 47).
- Count down from 64 to 28 subtracting 6 (i.e., 64, 58, 52, 46, 40, 34, and 28).
- Count down from 76 to 27 subtracting 7 (i.e., 76, 69, 62, 55, 48, 41, 34, and 27).

HOW TO SOLVE

Problem: 58 - 37 = ?

Solution: Step 1: 37 is made of 30 and 7.

Step 2: First, 58 - 30 = 28; then 28 - 7 = 21.

The answer: 58 - 37 = 21.

EXERCISE II

100 - 15 = 85	100 - 14 = 86	100 - 34 = 66	100 - 11 = 89
100 - 16 = 84	100 - 53 = 47	100 - 24 = 76	100 - 29 = 71
100 - 18 = 82	100 - 15 = 85	100 - 33 = 67	100 - 19 = 81
100 - 25 = 75	100 - 16 = 84	100 - 15 = 85	100 - 18 = 82
100 - 11 = 89	100 - 44 = 56	100 - 20 = 80	100 - 32 = 68
100 - 23 = 77	100 - 21 = 79	100 - 14 = 86	100 - 50 = 50
100 - 16 = 84	100 - 29 = 71	100 - 13 = 87	100 - 25 = 75
100 - 17 = 83	100 - 26 = 74	100 - 27 = 73	100 - 90 = 10
100 - 43 = 57	100 - 27 = 73	100 - 25 = 75	

EXERCISE III

70 - 56 = 14	50 - 33 = 17	85 - 75 = 10	90 - 18 = 72
80 - 64 = 16	50 - 36 = 14	50 - 18 = 32	80 - 15 = 65
90 - 67 = 23	60 - 42 = 18	60 - 13 = 47	80 - 28 = 52
50 - 36 = 14	70 - 43 = 27	70 - 18 = 52	90 - 28 = 62
60 - 22 = 38	80 - 72 = 8	80 - 43 = 37	50 - 23 = 27
70 - 45 = 25	60 - 33 = 27	90 - 28 = 62	40 - 17 = 23

WORD PROBLEMS

1. Three numbers add to up 100. Two of them are 20 and 50. What is the third number? *Ans:* 30.
    Solution: Two numbers together make 20 + 50 = 70. Now, 70 and the third number together make 100. Then, 100 - 70 (two number together) = 30 (the third number).

2. Three numbers add to up 100. Two of them are 70 and 30. What is the third number? *Ans:* 0.
    Solution: Two numbers together make 70 + 30 = 100. Now, 100 and the third number together make 100. Then, 100 - 100 (two number together) = 0 (the third number).

3. A stock was $75 but then overnight it zoomed to $100. By how much did its price increase in a single day? *Ans:* $25.

4. Nick bought a skateboard for $67 and later sold it for $41. How much money did he lose? *Ans:* $26.

5. A barrel held 48 gallons of water but 22 gallons leaked out. How much water remained in the barrel? *Ans:* 26 gallons.

6. Uncle Luis is 57 years old.
   a) He is 24 years older than Cousin Fay. How old is Fay? *Ans:* 33 years old.
   Solution: Uncle Luis is older. That means he is 24 years older than Fay. If he is 57 years old, she is 57 - 24 = 33 (years old).
   b) He is 24 years younger than Aunt Rose. How old is Aunt Rose? 81 years old.
   Second Solution: Uncle Luis is younger. That means he is 24 less years than Aunt Rose. If he is 57 years old, she is 57 + 24 = 81 (years old).
   c) He is 11 years younger than Uncle Larry. How old is Uncle Larry? *Ans:* 68 years old.
   d) He is 11 years older than Cousin Tom. How old is Cousin Tom? *Ans:* 46 years old.

7. A boat rental shop has 45 boats to rent out. It rented 14 boats before 10 o'clock. How many boats are still available? *Ans:* 31 boats.

8. Bill threw a ball up in the sky 56 times and said "Oops" 32 times. How many times did the ball not land on his head? *Ans:* 24 times.

9. Holly is on page 37 of a 59-page story. How many more pages are left to the end? *Ans:* 22 pages

10. Out of 68 days of summer vacation, Mr. Rossi traveled for 43 days. How many days did he stay home? *Ans:* 25 days.

11. An artist drew 35 pictures and threw away 24. How many did she keep? *Ans:* 11 pictures.

12. Troy broke rules 48 times and apologized 33 times. How many more apologies are needed? *Ans:* 15 apologies.

13. Morgan put 48 blueberries in her cereal but only 25 grapes. How many more berries than grapes did she eat with her cereal? *Ans:* 23 blueberries.

14. Ms. Cozy bought 87 feet of wall paper and used 64 feet in one room. How many feet are left for the other room? *Ans:* 23 feet.

15. Out of 55 miles of a road, only 33 are paved. How many miles are unpaved? *Ans:* 22 miles.

16. Out of 100 children at the game, 25 were girls. How many boys were there? *Ans:* 75 boys.

17. In a 100-meter dash, Nick ran 45 meters and then tripped. How much was left to run before he fell? *Ans:* 55 meters.

18. Out of 58 taxi cabs at the station, 24 need repairs. How many are working okay? *Ans:* 34 taxi cabs.

19. Out of 68 watches in a store 54 were wrist and the rest were pocket watches. How many pocket watches were in the store? *Ans:* 14 pocket watches.

20. From a 64 oz. bottle of milk, Amita poured out 32 oz. How many ounces are left in the bottle? *Ans:* 32 oz. She poured out exactly one half.

21. Mrs. Tracy is 68.
    a) How old was she 13 years ago? *Ans:* 55 years old.
    b) How old was she 25 years ago? *Ans:* 43 years old.
    c) How old was she 37 years ago? *Ans:* 31 years old.
    d) How old was she 44 years ago? *Ans:* 24 years old.

22. A store received 49 very popular toy cars. They sold 17 cars in the first hour. How many cars are left? *Ans:* 32 cars.
    a) How many cars will the store have left if it sells 25?
    Ans: 24 cars.
    b) How many cars will the store have left if it sells 33?
    Ans: 16 cars.
    c) How many cars will the store have left if it sells 39?
    Ans: 10 cars.

23. Alice bought 50 chestnuts and roasted 25 today. How many will she roast next time? *Ans:* 25 chestnuts. Did you remember that one half of 50 is 25?

24. Carmen paid $48 for a lamp at a sale. Before the sale, the lamp was $60. How much money did she save? *Ans:* $12.

25. Naomi paid for a $33 phone bill with a $50 bill. How much change did she get back? *Ans:* $17.

26. A man owns 70 acres of land and planted lettuce on 38 acres and tomatoes on the rest. How much land did he plant with tomatoes? *Ans:* 32 acres.

27. What number plus 36 equal 60? *Ans:* 24.

28. What number plus 45 equal 80? *Ans:* 35.

29. What number plus 57 equal 90? *Ans:* 33.

30. Once upon a time there was a man who found 100 golden coins in the forest. The man took the 100 golden coins and started for home.
a) On the way home he saw a baker and gave him 33 coins for 3 pies. How many coins did he have left? *Ans:* 67 coins.
b) Then he met a neighbor who asked to borrow 28 coins. How much did the man have then? *Ans:* 39 coins.
c) While crossing a stream, the man stumbled on a rock and dropped 38 coins in the river. What did he bring home? *Ans:* 1 coin and 3 pies. May be he dropped the pies in the river too!

31. A hot air balloon went up 78 meters and then settled down 25 meters. How high from the ground is the balloon? *Ans:* 53 meters.

32. Maggie thought the books would cost $42 but they cost $67 instead. How much more did the books cost than what she was expecting to pay? *Ans:* $25.

33. A hall has 62 seats but 88 people came to listen to the famous professor. How many will have to stand? *Ans:* 26 people.

34. Carl sent out 56 invitations and asked for RSVP. He got back response back from only 31 people. How many have still to RSVP? *Ans:* 25 people. What is RSVP?

35. It took 87 days to build a house, during which time the builders took 16 days off. How many days did they actually work? *Ans:* 71 days.

36. There are 96 feet from the tip of the tree on a hill to the sea level. The tree is 24 feet, how high is the hill? *Ans:* 72 feet.

37. A baseball player swung 69 times and missed 42. How many times did he hit? *Ans:* 27 times.

38. A teacher had 45 pencils and gave away 25 at an exam. How many pencils did she keep? *Ans:* 20 pencils.

39. A yellow ribbon is 60 inches long. A green ribbon is 17 inches shorter. How long is the green ribbon? *Ans:* 43 inches.

40. When Josh finished the 100-meter dash, Kim was 16 meters behind. How far from the start was Kim? *Ans:* 84 meters.

41. A 100-minute show included 23 minutes of intermission. How long was the performance? *Ans:* 77 minutes.

42. The eldest princess has 80 pairs of shoes; the younger princess has 26 pairs less. How many pairs does the younger princess have? *Ans:* 54 pairs.

43. A truck with 50 TV sets delivered 19 to a store. How many sets are on the truck now? *Ans:* 31 TVs.

44. From a set of 70 building blocks, Cooper took 41 to build a house. How many blocks he did not use? *Ans:* 29 blocks.

45. On the American flag 50 stars stand for 50 states. The first flag had only 13 stars. How many stars were added later?
    Ans: 37 stars.

46. McKenzie got a book with 40 mazes and successfully traced 24 of them. How many were too hard to trace? *Ans:* 16 mazes.

47. Wilhelm Tell practiced archery with apples. Out of 70 shots, his arrow hit apples 58 times. How many times did he miss?
    Ans: 12 times.

48. Inside a piñata there were 80 candies. 42 fell out after the first blow. How many are still inside? *Ans:* 38 candies.

49. A centipede had 100 legs. He tried to ride a skateboard and broke 25 of his legs. He still has how many good legs?
    Ans: 75 legs.

50. A store sold 70 bars of chocolate; 22 were milk chocolate, the rest were dark chocolate? How many bars of dark chocolate did the store sell? *Ans:* 48 bars.

51. There were 60 days left before the New Year's Day. How many days were left to the end of the year after 2 weeks passed? **Ans:** 46 days.
Solution: A week is 7 days. There are 14 days in 2 weeks. Then, 60 - 14 = 46 (days).

52. A veterinarian saw 50 pets in one day. Of these 14 were birds and 3 were gold fish and the rest were cats and dogs. How many cats and dogs did she see? **Ans:** 33 cats and dogs (14 + 3 = 17, then 50 - 17 = 33).

53. At a fun park, Keesha took 40 rides. She closed her eyes on 18 rides. On how many rides did she keep her eyes open? **Ans:** 22 rides. Why do you think she kept her eyes closed?

54. Every night for 100 days, Mr. Fret had dreams. He counted his dreams and found that he had 39 good dreams, 12 bad dreams and the rest he could not remember. How many dreams did Mr. Fret forget all about? **Ans:** 49 dreams.

55. Ingrid knew the capitals of all 50 states but 11. How many capitals can she remember? **Ans:** 39 capitals.

56. Tyrone put $90 in the bank and then took out $29. How much money is in account now? **Ans:** $61.

57. Ingrid knew the capitals of all 50 states but 11. How many capitals can she remember? **Ans:** 39 capitals.

58. Out of 31 days this month, it was raining for 12 days. How many days it was not raining? **Ans:** 19 days.

59. Nineteen days this month it was not raining, and was sunny on 8 days. How many days it was not sunny? **Ans:** 23 days. Ha, tricked you!

60. Out of 31 days this month, it was raining for 12 days and foggy for 4 days. How many days it was not foggy or raining? **Ans:** 15 days.

17

SUBTRACTING DOUBLE-DIGIT NUMBERS

EXERCISE I

- Count from 0 to 100 by 10 (i.e., 0, 10, 20, 30, etc.)
- Count from 100 back to 0 by tens.

HOW TO SOLVE

First Problem: 57 - 48 = ?

Solution: If the problem was 57 - 47, then it would be easy to subtract, because the answer is 10. In this example we can use a trick. In our minds we can take away 1 from 48 and put in our "memory bank". Now we have 47 and can take away 47 from 57. Easy: 57 - 47 = 10. But we made the number we subtracted "lighter" be taking away 1. So, we take 1 from the "memory bank" and take it away from 10: 10 - 1 = 9.

The answer: 57 - 48 = 9.

It is always a good idea to recheck the answer.

Second Problem: 67 - 19 = ?

Solution: In this problem it might be easier to add 1 to number 19 and make it 20. Then 67 - 20 = 47, easy! But we took away more, a number "heavier" by 1 and so need to correct it by giving it back: 47 + 1 = 48

The answer: 67 - 10 = 48.

It is very important not to get confused when we take away or add the digits from the "memory bank".

EXERCISE II

34 - 10 = 24	38 - 22 = 16	53 - 36 = 17	57 - 18 = 39
34 - 21 = 13	39 - 25 = 14	55 - 42 = 13	47 - 33 = 14
35 - 25 = 10	43 - 19 = 24	59 - 43 = 16	47 - 28 = 19
35 - 15 = 20	45 - 19 = 26	58 - 12 = 46	57 - 18 = 39
36 - 22 = 14	45 - 26 = 19	57 - 17 = 40	58 - 15 = 43
36 - 17 = 19	47 - 22 = 25	67 - 15 = 52	59 - 28 = 31
37 - 16 = 21	49 - 35 = 14	68 - 18 = 50	49 - 13 = 36
37 - 18 = 19	49 - 33 = 16	67 - 13 = 54	64 - 32 = 32

WORD PROBLEMS

1. In a theater that can seat 62, 43 people attended a show. How many seats were empty? *Ans:* 19 seats.

2. There were 43 people in the theater at the beginning of the show. By the end there were 14. How many people left early? *Ans:* 29 people.

3. Mr. Butler planned a party for 37 guests but 46 people showed up. How many extra guests came? *Ans:* 9 guests.

4. Mrs. Butler planted 41 bulbs but only 25 bulbs grew. How many bulbs didn't make it? *Ans:* 16 bulbs.

5. School's debate team had 44 members and then they added 19 more. How many students are on the team now? *Ans:* 63 members.

6. If you hold 53 beads in two fists and the right fist holds 27 beads, how many are in the left? *Ans:* 26 beads.

7. Lisa earned $64 doing baby sitting and spent $35 on books. How much did she save? *Ans:* $29.

8. A man was 25 when he got married. He is now 53. How long has he been married? *Ans:* 28 years.

9. Nolan bought 45 cookies but brought home only 39. How many cookies didn't make it home? *Ans:* 6 cookies.

10. There are 70 nails in a box; 36 short and the rest are long. How many long nails are in the box? *Ans:* 34 nails.

11. Wendy's mom is 37 years old and her dad is 45 years old. By how much is her dad older than her mom? *Ans:* 8 years.

12. Michael won 44 chess games, Jorge won 15 less.
    a) How many games did Jorge win? *Ans:* 29 games.
    b) How many games did they both win? *Ans:* 73 games.

13. One watch costs $53. The second watch cost $26 less.
    a) How much does the second watch cost? *Ans:* $27.
    b) How much do both watches cost? *Ans:* $80.

14. During summer, Tammy read 46 books. Sonny read 18 books fewer.
    a) How many books did Sonny read? *Ans:* 28 books.
    b) How many books did they both read? *Ans:* 74 books.

15. Farmer Ted has 34 sheep; farmer Joe has 3 less. How many sheep do they both have? *Ans:* 65 sheep.

16. Farmer Ted has 27 goats; farmer Joe has 5 more. How many goats do they both have? *Ans:* 59 goats.

17. Framer Ted has 28 cows; farmer Joe has 6 more. How many cows do they both have? *Ans:* 62 cows.

18. A professor took 62 books from the library and returned 39 the same day. How many books did he keep to read?
    Ans: 23 books.

19. One tree is 77 feet tall; the other is 19 feet shorter. How tall is the other tree? *Ans:* 58 feet.

20. An artist sketched 63 pictures in pencil and painted only 27 of them. How many sketches were kept as sketches? *Ans:* 36.

21. A 56 bed hospital has 19 patients. How many more patients can the hospital admit? *Ans:* 37 patients.

22. A basketball game ended with a score of 61 to 58. How many points separate the two teams? *Ans:* 3 points.

23. There are 55 water lilies in the pond; 37 frogs sat, one each, on a lily. How many empty lilies are in the pond? *Ans:* 18 lilies.

24. Out of 46 pieces of cheese left overnight on the table, mice ate 17. How many pieces were spared? *Ans:* 29 pieces.

25. Murray paid $72 for a fur hat and a scarf. The scarf cost $27. How much was the hat? *Ans:* $45.

26. A team bought 51 hockey sticks but had only 14 by the end of the season. How many sticks did they break during the season? *Ans:* 37 sticks.

27. Eighty-two invitations were sent and 66 people came to party. How many didn't? *Ans:* 16 people.

28. If out of 43 french fries on my plate I dipped 19 in ketchup, how many I did not dip? *Ans:* 24 fries.

29. If out of 64 shirts in the laundry 39 shirts were starched, how many were not? *Ans:* 25 shirts.

30. If out of 72 fortunes from fortune cookies, 18 came true, how many didn't? *Ans:* 54 fortunes.

31. If out of 58 rooms in a hotel, 39 rooms are occupied, how many rooms are empty? *Ans:* 19 rooms.

32. There are 44 students in my class; 25 are girls. How many boys are in the class? *Ans:* 19 boys.

33. We bought 65 cans of paint; 26 cans are brown paint, the rest are yellow. How many yellow paint cans did we buy? *Ans:* 39 cans.

34. Seventy-four migrating birds landed on a lake. There are 37 geese and the rest are ducks. How many ducks are on the lake? *Ans:* 37 ducks.

35. A construction site has 43 safety hats and requires all workers to wear a hat while working.
    a) If there are only 16 hats on the shelf, how many people are working today? *Ans:* 27 workers.
    b) If 25 hats out of all Safety hats are yellow and the rest are red, how many red hats do they have? *Ans:* 18 red hats.

36. Altogether 81 finches and chickadees live in the park. If there are 36 chickadees, how many finches live there? *Ans:* 45 finches.

37. There were 73 cars on the parking lot in the morning, and now there are only 25. How many cars left the lot? *Ans:* 48 cars.

38. A restaurant received reservations for 64 people but only 55 people showed up. How many didn't come? *Ans:* 9 people.

39. For the party Bobbie made 53 glasses of lemonade and 16 glasses of apple juice. How many drinks did she make? *Ans:* 69 glasses.

40. A train had 52 cars. At the station, 13 cars were removed. How many cars remained? *Ans:* 39 cars.

41. At night 58 animals came to a small watering hole. There were 19 zebras and the rest were gazelles. How many gazelles came to drink water? *Ans:* 39 gazelles.

42. During the summer, the weather service counted 66 tornadoes. Of these 11 tornadoes caused damage. How many didn't? *Ans:* 55 tornadoes.

43. Out of 90 days of summer, Blair watered the lawn 21 days. On how many days did he not water the lawn? *Ans:* 69 days.

44. In a speech given by Mr. French, 11 out of 46 words were in French. How many words are not? *Ans:* 35 words.

45. A doctor looked at 72 X-rays and found problems in 27 of them. How many X-rays were okay? *Ans:* 45 X-rays.

46. A farmer had to cut 18 out of 82 trees to make a road through the woods.
a) How many trees were not cut? *Ans:* 64 trees.
b) Then the farmer planted 27 trees at the edge of the new road. How many trees are there now? *Ans:* 91 trees.

47. A model tried on 39 dresses. How many dresses she did she not try, if there were 98 dresses in the store? *Ans:* 59 dresses.

48. From the top of his roof, Mr. Gonzales can see 42 out of 73 houses in his village. How many houses can he not see?
Ans: 31 houses.

49. On the test, Cory got the right answer 37 times and wrote "I don't know" 24 times. How many questions were on the test?
Ans: 61 questions.

50. On the way to the lake, Luis yawned 33 times and Leigh yawned 14 times less. How many times did Leigh yawn?
Ans: 19 times.

51. The distance between New York City and Philadelphia is 97 miles . How much distance remains after driving the first 49 miles? *Ans:* 38 miles. (97 - 40 = 57; 57 - 9 = 48).

52. Jackie invited 4 friends to dinner. They each brought one of their friends. How many people came to the dinner?
Ans: 8 people.

53. After the dinner, 3 couples showed up. How many people are the dinner now? *Ans:* 14 people.

54. Jackie had prepared 11 pieces of cake for the dinner, how many people will not get cake if there are 19 people now at the dinner party? *Ans:* 8 people.

55. Jackie opened a box of cookies for the guests. There were only 9 cookies there. She needs twice as many. How many cookies does she need for the guests? *Ans:* 1 cookies.

18

STARTING MULTIPLICATION

Multiplication is repeated addition. If we add 2 + 2 + 2 = 6, it means we add 2 three times. We can also say 2 times 3 equals 6. That becomes even more important when we add many of the same numbers. For example, 2 + 2 + 2 + 2 + 2 + 2 + 2 + 2 = 14 also means that we added 2 seven times. Saying that we multiplied 2 by 7 times is saying the same thing. 2 × (times) 7 = 14.

Another way to understand multiplication is to think of it as addition of sets of numbers. If you imagine one set (a set is like a mental basket) of 4 jelly beans, then two sets (baskets) which will have 8 beans, and 3 baskets will have 12 beans. In mathematics we say that we take one set (4 beans) 2 times or 3 times, or 10 times or even 100 times (lots of beans, right?).

We can say that 3 times 2 equals 6 meaning that a set of 3 was counted twice (like this: 3 + 3) and the answer is 6. Try it.

- 2 sets of 3 pebbles equals 6 pebbles.

- 2 sets of 4 sea shells equals 8 sea shells.

- 2 sets of 5 toes equals 10 toes.

- 3 sets of 3 pillows equals 9 pillows.

- 3 sets of 10 nails equals 30 nails.
- 2 sets of 12 eggs equals 24 eggs.

To help with the concept of multiplication, use some dry beans or marbles. Ask child to place them in 3 groups of 2 beans each on the table. Count aloud the number of beans and say: "In math we say if we take 2 beans three times we will have 6 beans. Or 2 times 3 makes 6." Regroup the same 6 beans into 2 groups of 3 beans each and say: "Now we take 3 beans 2 times and that also makes 6. Or 3 times 2 makes 6." Do it again with a different set of beans, e.g. 3 sets of 4 beans, 3 sets of 3 beans, 5 sets of 2 beans, on so on.

EXERCISE I

- 2×2 is the same as $2 + 2$ and it equals 4.
- 4×2 is the same as $4 + 4$ and it equals 8.
- 2×3 is the same as $2 + 2 + 2$ and it equals 6.
- 5×2 is the same as $5 + 5$ and it equals 10.
- 3×3 is the same as $3 + 3 + 3$ and it equals 9.
- 4×3 is the same as $4 + 4 + 4$ and it equals 12.
- 4×4 is the same as $4 + 4 + 4 + 4$ and it equals 16.
- Count from 3 to 50 skipping 3 numbers; or, in other words, by adding 4. (3, 7, 11, 15, 19, 23, 27, etc.)

EXERCISE II

I will count by 2's but skip a number. Tell me what number I skipped.

- 2, 6, 8   What is missing? *Ans:* 4
- 2, 4, 8   What is missing? *Ans:* 6
- 4, 6, 8   What is missing? *Ans:* Nothing is missing
- 6, 10, 12   What is missing? *Ans:* 8
- 12, 16, 18   What is missing? *Ans:* 14
- 10, 12, 16   What is missing? *Ans:* 14

- 2, 6, 8   What is missing? *Ans:* 4
- 4, 6, 10   What is missing? *Ans:* 8
- 14, 16, 20   What is missing? *Ans:* 18
- 20, 22, 26   What is missing? *Ans:* 24
- 84, 88, 90   What is missing? *Ans:* 86

EXERCISE III

$2 \times 2 = 4$	$2 \times 9 = 18$	$5 \times 2 = 10$	$6 \times 2 = 12$
$2 \times 1 = 2$	$5 \times 2 = 10$	$7 \times 2 = 14$	$8 \times 2 = 16$
$3 \times 2 = 6$	$6 \times 2 = 12$	$2 \times 4 = 8$	$10 \times 1 = 10$
$2 \times 4 = 8$	$2 \times 7 = 14$	$2 \times 8 = 16$	$2 \times 10 = 20$
$4 \times 2 = 8$	$8 \times 2 = 16$	$5 \times 2 = 10$	$4 \times 2 = 8$
$2 \times 5 = 10$	$2 \times 3 = 6$	$2 \times 9 = 18$	$2 \times 7 = 14$

WORD PROBLEMS

1. How many tails do 3 blind mice have? *Ans:* 3 tails.

2. If one tile costs $2, how much do 2 tiles cost? *Ans:* $4.

3. If one hamster is $7, how much will you pay for 2 hamsters? *Ans:* $14.

4. In the library, each child can take out 5 books. How many books can 2 children take out? *Ans:* 10 books.

5. If there are 4 guitar players, how many guitars do they have if each has only one? *Ans:* 4 guitars, because $4 \times 1 = 4$.

6. How many shoes do 4 guitar players wear? *Ans:* 8 shoes ($2 \times 4 = 8$).

7. If the guitar player has 3 rings in each ear, how many earrings does he wear? *Ans:* 6 rings ($3 \times 2 = 6$).

8. Three kids have 2 books each. How many books do they have together?

9. If 2 kids have 5 apples each, how many apples do they both have? *Ans:* 10 apples.

10. If one unicorn has one horn, how many horns do 7 unicorns have? *Ans:* 7 horns.
    Unicorn is an imaginary animal that looks like a white deer/horse with a horn in the middle of his forehead.

11. We have 8 kids on our gymnastics team and the other team has 2 times as many. How many kids do they have on their team? *Ans:* 16 kids.

12. If one plant has 2 seeds, how many seeds do 4 plants have? *Ans:* 8 seeds.

13. If one cricket can eat 7 seeds, how many seeds can 2 crickets eat? *Ans:* 14 seeds.

14. If one lizard can eat 7 crickets, how many crickets can 2 lizards eat? *Ans:* 14 crickets.

15. If two lizards can eat 8 crickets, how many crickets can eat 2 lizards? Ha, gotcha. Crickets can't eat lizards!

16. If one lizard costs $8, how much do 2 lizards cost? *Ans:* $16.

17. If one vulture has 2 legs, how many legs do 9 vultures have? *Ans:* 18 legs.

18. If one tarantula has 8 legs, how many legs do 2 of them have? *Ans:* 16 legs.

19. How many noses do 5 boys have? *Ans:* 5 noses.
    a) How many noses do 10 boys have? *Ans:* 10 noses.
    b) How many noses do 13 boys have? *Ans:* 13 noses.

20. If 5 apples cost $1, how many can you buy for $2? *Ans:* 10 apples.

21. If there are 4 speakers in each box, how many speakers are in 2 boxes? *Ans:* 8 speakers.

22. In an airplane, 6 passengers can sit in each row. How many passengers can sit in 2 rows? *Ans:* 12 passengers. In a big airplane, how many passengers usually sit in a row? As many as 10.

23. If a bicycle has 2 pedals, how many pedals do 10 bicycles have? *Ans:* 20 pedals.

24. If a unicycle has 2 pedals, how many pedals do 10 unicycles have? *Ans:* Also 20 pedals.

25. One elephant has 2 tusks.
    a) How many tusks do 3 elephants have? *Ans:* 6 tusks.
    b) How many tusks do 5 elephants have? *Ans:* 10 tusks.
    c) How many tusks do 7 elephants have? *Ans:* 14 tusks.

26. A car has 4 wheels, a scooter has 2. How many wheels do they both have? *Ans:* 6 wheels (4 + 2 = 6). This is an addition, not a multiplication problem.

27. A triangle has 3 sides. How many sides do 2 triangles have? *Ans:* 6 sides. Ask child to draw a triangle.

28. A square has 4 sides. How many sides do 2 squares have? *Ans:* 8 sides. Ask child to draw a square.

29. A pentagon has 5 sides. How many sides do 2 pentagons have? *Ans:* 10 sides. Penta means five. Ask child to draw a pentagon.

30. A hexagon has 6 sides. How many sides do 2 hexagons have? *Ans:* 12 sides. Hex means 6. Ask child to draw a hexagon.

31. Jill runs 7 miles every day. How many miles would she run in 2 days? *Ans:* 14 miles.

32. Raja runs 5 miles every day. How many miles would he run in 2 days? *Ans:* 10 miles.

33. Mr. Arnold can work on a project either 2 hours each day for 8 days, or 8 hours each day for 2 days. Which way would he work more hours? *Ans:* Either way it makes the same 16 hours (2×8 = 16 and also $8 \times 2 = 16$).

34. A barber can shave 4 beards in an hour. How many beards can the barber shave in 2 hours? *Ans:* 8 beards.

35. Another barber can trim 7 mustaches in 1 hour. How many mustaches can this barber trim in 2 hours? *Ans:* 14 mustaches.

36. Yet another barber can cut hair for 2 customers in 1 hour. For how many customers can he cut hair in 4 hours? *Ans:* 8 customers.

37. A car mechanic changes oil in 9 cars in one hour. How many oil changes can he do in 2 hours? *Ans:* 18 oil changes.

38. The mechanic can replace 2 brakes in one hour. How many brakes can he replace in 2 hours? *Ans:* 4 brakes.

39. If each car has four brakes to be replaced, and the mechanic can replace 2 brakes in one hour, how long would it take him to replace brakes in two cars? *Ans:* 4 hours. This one may be hard for the child.

40. The mechanic can paint 2 cars a day. How many cars can he paint in 9 days? *Ans:* 18 cars.

41. The same mechanic can paint 2 cars in 1 day. How many motorcycles can he paint in 6 days? *Ans:* We don't know! We only know about the cars, not motorcycles.

42. A pair means 2 of anything.
    a) How many bears are in 5 pairs of bears? *Ans:* 10 bears.
    b) How many pears are in 8 pairs of pears? *Ans:* 16 pears.
    c) How many chairs are in 9 pairs of chairs? *Ans:* 18 chairs.
    d) How many hares are in 7 pairs of hares? *Ans:* 14 hares.
    e) How many squares are in 10 pairs of squares?
    Ans: 20 squares.

43. An ice cream sundae costs $4. How much do 2 sundaes cost? *Ans:* $8.

44. Nelly spent $10 for a birthday gift for her friend. How much would she spend for gifts if her friend had a twin? *Ans:* $20.

45. The ancient world had 7 wonders. How many wonders would have been in 2 ancient worlds? *Ans:* 14 wonders.

46. If one cell phone battery lasts 9 hours, how long will 2 batteries last? *Ans:* 18 hours.

47. If one set of chopsticks has 2 sticks, how many sticks are in 2 sets? *Ans:* 4 chopsticks.

48. If there are 6 oars on one side of the boat, how many oars are on both sides? *Ans:* 12 oars.

49. If there are 9 windows on one side of the airplane, how many are on both sides? *Ans:* 18 windows.

50. If there are 5 earrings in a teenager's right ear, how many rings are in both ears? *Ans:* Hard to tell, teenagers can be very confusing, but if you guessed 10, you are probably right.

51. A waiter put on the table, 7 glasses of Shirley Temples, with 2 cherries in each glass. How many cherries are in all drinks? *Ans:* 14 cherries.

52. If one lucky clover has 4 leaves, how many leaves are on 2 lucky clovers? *Ans:* 8 leaves.

53. If 2 eggplants cost $1, how many can you buy for $5? *Ans:* 10 eggplants.

19

MULTIPLICATION BY 2 OR 3

A pair means two of something. A pair of shoes means 2 shoes, a pair of socks means 2 socks, and a pair of monkeys means 2 monkeys. The exceptions are a pair of glasses, pair of scissors, and a pair of pants. Can you think of any other exceptions?

Remember that multiplication is reversible: 2 puppies in 4 baskets or 4 puppies in 2 baskets both will be 8 puppies. It works the reverse way too.

Now, let's learn the times table for 3. Try to memorize it and test yourself several times by doing problems in order and randomly.

EXERCISE I

$3 \times 1 = 3$	$3 \times 8 = 24$	$5 \times 3 = 15$	$3 \times 4 = 12$
$3 \times 2 = 6$	$3 \times 9 = 27$	$6 \times 3 = 18$	$4 \times 4 = 16$
$3 \times 3 = 9$	$3 \times 7 = 21$	$7 \times 3 = 21$	$4 \times 5 = 20$
$3 \times 4 = 12$	$1 \times 3 = 3$	$8 \times 3 = 24$	$3 \times 3 = 9$
$3 \times 5 = 15$	$2 \times 3 = 6$	$9 \times 3 = 27$	
$3 \times 6 = 18$	$3 \times 3 = 9$	$10 \times 3 = 30$	
$3 \times 7 = 21$	$4 \times 3 = 12$	$2 \times 3 = 6$	

EXERCISE II

2 × 6 = 12	3 × 5 = 15	8 × 2 = 16	2 × 5 = 10
4 × 2 = 8	3 × 4 = 12	9 × 2 = 18	2 × 6 = 12
3 × 3 = 9	2 × 9 = 18	8 × 3 = 24	2 × 5 = 10
3 × 2 = 6	6 × 3 = 18	3 × 9 = 27	2 × 10 = 20
3 × 4 = 12	7 × 3 = 21	3 × 8 = 24	
5 × 3 = 15	2 × 5 = 10	2 × 2 = 4	
2 × 7 = 14	6 × 3 = 18	2 × 4 = 8	

In math we say that there are as many, or 2 times as many, or 3 times as many of something than the other. For example: There are as many noses as there are children in the room, because if there are 5 children in the room and each person has one nose, then 5 × 1 = 5 noses.

There will be 2 times (or twice) as many ears as there are donkeys because every donkey has two ears, but there will be 4 times as many legs because each donkey has 4 legs.

There will be twice as many antennas as the number of ladybugs, because each ladybug has two, but there will be six times as many legs as the number of lady bugs, because each lady bug has six legs.

If one team scored 2 goals and the other team scored 4, we can also say that the other team scored twice as many goals, because we know that 2 × 2 = 4.

EXERCISE III

1. How many times more wings are there than chickens?
   Ans: 2 times more, because each chicken has 2 wings.

2. How many times more running shoes are there than runners?
   Ans: 2 times more, because each runner has two shoes.

3. How many times more tires are there than cars?
   Ans: 4 times more, because each car has 4 tires.

4. How many times more necks are there than people?
   Ans: the same number of times, because each person has only one neck.

5. How many times more wheels are there than tricycles?
   Ans: 3 times more, because each tricycle has 3 wheels.

6. How many times more toes are there than people?
   Ans: 10 times more, because each person has 10 toes.

7. How many times more days of the week are there than weeks?
   Ans: 7 times more, because each week has 7 days.

8. How many times more wheels are there than wheelbarrows?
   Ans: 1 times more (usually). Wheelbarrows can also have three or four wheels.

9. How many times more spider legs are there than spiders?
   Ans: 8 times more.

10. How many times more sides are there than triangles?
    Ans: 3 times more.

11. How many times more tentacles are there than octopi?
    Ans: 8 times more.

WORD PROBLEMS

1. Two desk lamps came in one box
   a) How many lamps are in 2 boxes? *Ans:* 4 lamps.
   b) How many lamps are in 3 boxes? *Ans:* 6 lamps.

2. If there are 2 students at each desk and there are 3 desks, how many students are there? *Ans:* 6 students.

3. Sonia has 3 chickens and each gave 2 eggs. How many eggs did she get? *Ans:* 6 eggs.

4. Kim has 2 chickens and each gave 3 eggs. How many eggs did Kim get? *Ans:* 6 eggs.

5. Kishore has 3 chickens and each gave 3 eggs. How many eggs did he get? *Ans:* 9 eggs.

6. There are 3 pairs on the dance floor. How many people are dancing? *Ans:* 6 people.

7. When 2 friends bought 3 pencils each, how many pencils did they both buy? *Ans:* 6 pencils.

8. I saw 4 girls wearing earrings. How many earrings were girls wearing?
   Ans: 8 earrings. This is probably a correct answer, although, some girls do wear more than two earrings and also just one earring as well.

9. If I see 3 plates with 3 plums on each plate. How many plums do I see? *Ans:* 9 plums.

10. In storage we found 2 boxes. When we opened the boxes, we found each box had 4 cans of paint. How many cans of paint did we find? *Ans:* 8 cans.

11. Two houses are next to each other.
    a) Each house has 2 chimneys. How many chimneys are in both? *Ans:* 4 chimneys.
    b) Each house has 3 doors. How many doors do they all have? *Ans:* 6 doors.
    c) Each house has 5 windows. How many windows do they all have? *Ans:* 10 windows.
    d) Each house has 1 guard dog. How many dogs are in all? *Ans:* 2 dogs.

12. LuAnn bought 5 notebooks at $2 each. How much did she spend? *Ans:* $10.

13. Joann bought 2 binders at $5 each. How much did she spend? *Ans:* $10.

14. Tim saw 6 nests, and each nest had 2 eggs. How many eggs did he see? *Ans:* 12 eggs.

15. Sam saw 2 nests with 6 eggs in each. How many eggs did she see? *Ans:* 12 eggs.

16. A rowboat has 6 oars on each side. How many oars does it have? *Ans:* 12 oars.

17. How many legs do 3 little pigs have? *Ans:* 12 legs.

18. How many eyes do 3 little pigs have? *Ans:* 6 eyes.

19. Four kids had breakfast. Each ate 3 toasts. How many toasts did they all eat? *Ans:* 12 toasts.

20. Nadia wrote some words on a page. Each of these words had 3 letters.
    a) How many letters are in 3 words? *Ans:* 9 letters.
    b) How many letters are in 4 words? *Ans:* 12 letters.
    c) How many letters are in 2 words? *Ans:* 6 letters.

21. If a nickel is worth 5 cents,
    a) How many cents are in 2 nickels? *Ans:* 10 cents.
    b) How many cents are in 3 nickels? *Ans:* 15 cents.

22. One hand has 5 fingers.
    a) How many fingers do 2 hands have? *Ans:* 10 fingers.
    b) How many fingers do 3 hands have? *Ans:* 15 fingers.

23. Two chandeliers hang in the hall. Each has 7 light bulbs.
    a) How many light bulbs do they both have?
    Ans: 14 light bulbs.
    b) How many light bulbs would be in 3 chandeliers?
    Ans: 21 light bulbs.

24. If one fox has 4 legs, how many legs do 3 foxes have?
    Ans: 12 legs.

25. One cockroach has 6 legs.
    a) How many legs do 2 cockroaches have? *Ans:* 12 legs.
    b) How many legs do 3 cockroaches have? *Ans:* 18 legs.

26. One octopus has 8 tentacles.
    a) How many tentacles do 2 octopi have? *Ans:* 16 tentacles.
    b) How many tentacles do 3 octopi have? *Ans:* 24 tentacles.

27. If one triangle has 3 sides,
    a) How many sides do 2 triangles have? *Ans:* 6 sides.
    b) How many sides do 3 triangles have? *Ans:* 9 sides.
    c) How many sides do 4 triangles have? *Ans:* 12 sides.

28. If one square has 4 sides,
    a) How many sides do 2 squares have? *Ans:* 8 sides.
    b) How many sides do 3 squares have? *Ans:* 12 sides.

29. One jacket has 3 pockets.
    a) How many pockets do 3 jackets have? *Ans:* 9 pockets.
    b) How many pockets do 4 jackets have? *Ans:* 12 pockets.
    c) How many pockets do 5 jackets have? *Ans:* 15 pockets.

30. How many legs do 3 dogs have? *Ans:* 12 legs.

31. How many legs do 4 cats have? *Ans:* 16 legs.

32. If one porter can carry 2 bags, how many bags can 6 porters carry? *Ans:* 12 bags.

33. How many ears do 7 rabbits have? *Ans:* 14 ears.

34. Two shelves have 6 cups each. How many cups are on both shelves? *Ans:* 12 cups.

35. How many legs do 2 spiders have if one spider has 8 legs? *Ans:* 16 legs.

36. How many arms do 8 princesses have? *Ans:* 16 arms.

37. How many legs do 3 cows have? *Ans:* 12 legs.

38. How many wings do 8 finches have? *Ans:* 16 wings.

39. How many noses do 8 ducks have? *Ans:* 8 noses.

40. A shirt has 8 buttons. How many buttons do 2 shirts have? *Ans:* 16 buttons.

41. Kim invited all her grandparents to a school concert and Lim also invited his 4 grandparents. How many grandparents did they both invite? *Ans:* 8 grandparents.

42. A star has 5 points.
    a) How many points do 2 stars have? *Ans:* 10 points.
    b) How many points do 3 stars have? *Ans:* 15 points.

43. A telephone has 3 buttons in each of the 4 rows. How many buttons does it have? *Ans:* 12 buttons.

44. Ann Marie used 3 roses in each bouquet to make 5 bouquets. How many roses did she use? *Ans:* 15 roses.

45. Mary Ann used 5 roses in each bouquet and made 3 bouquets. How many roses did she use? *Ans:* 15 roses.

46. Each floor in a 2-story building is 9 feet tall. How tall is the building? *Ans:* 18 feet.

47. How many legs do 3 flies have if one fly has 6? *Ans:* 18 legs.

48. A grand piano has 3 legs.
    a) How many legs do 6 pianos have? *Ans:* 18 legs.
    b) How many legs do 8 pianos have? *Ans:* 24 legs.
    c) How many legs do 7 pianos have? *Ans:* 21 legs.

49. If one cat has 9 lives,
    a) How many lives do 2 cats have? *Ans:* 18 lives.
    b) How many lives do 3 cats have? *Ans:* 27 lives.

50. How many wings do 10 eagles have? *Ans:* 20 wings.

51. One week has 7 days.
    a) How many days do 2 weeks have? *Ans:* 14 days.
    b) How many days do 3 weeks have? *Ans:* 21 days.
    c) How many days does a month have? *Ans:* 30 or 31.
    This may be a good time to discuss the knuckle rule for days in
    a month.

52. Mary had saved $12 and after one year she realized that she
    now had 3 times as much in her bank as before. How much
    money does she have in her bank? *Ans:* $36.

53. Mary had 5 red T-shirts but altogether has three times that
    many more. How many other T-shirts does Mary have?
    Ans: 15 T-shirts that are not red.

54. How many wings do 12 airplanes have? *Ans:* 24 wings.

55. How many wings do 6 boys have? *Ans:* None.

20

INTRODUCTION TO DIVISION

Division is easy – it's simply a reversal of multiplication. When multiplying, we add equal numbers or items into one group. When we divide we break a group or numbers into equal parts.

Let's start with division by 1 and by the number itself.
When we divide a number by 1 we try to answer 2 questions:
a) how many groups will be there if each group will have 1 in it?
b) what is the number in the group, if only 1 group is there?

$4 \div 1 = 4$
a) If we take 4 and divide it into groups of 1, how many groups will be there? *Ans:* 4 groups.
b) If we make 1 group, what number will be in the group? *Ans:* 4.

$12 \div 1 = 12$
a) If we take 12 and divide it into groups of 1, how many groups will be there? *Ans:* 12 groups.
b) If we make 1 group, what number will be in the group? *Ans:* 12.

What is:
- $68 \div 1 = 68$

- $100 \div 1 = 100$

- What is 2 divided by 1? *Ans:* 2
- What is 1 divided by 1? *Ans:* 1
- What is 100 divided by 1? *Ans:* 100
- What is 1,000 divided by 1? *Ans:* 1,000
- What is 1 million divided by 1? *Ans:* 1 million

When we divide a number by itself, mathematically thinking, we either:

a) Create as many groups as the number itself. If the number is 3 and we divide it by 3, then we divide 3 into 3 separate groups and each group has 1 in it.
b) Or we find out that if the entire number in the group is there only can be 1 group, because zeros do not make groups in arithmetic (and in life too).

Number 10 divided by 10 makes 1.

What is:
- $12 \div 12 = 1$
- $20 \div 20 = 1$
- $50 \div 50 = 1$
- $99 \div 99 = 1$
- $100 \div 100 = 1$
- 1 million divided by 1 million is 1.
- 1 gazillion $\div$ 1 gazillion also equals 1.

Now, let's take out the beans when we learned multiplication and use them now to learn division.

First, take 6 beans and divide them into 2 equal groups, 3 beans in each group. We divided 6 by 2 and came out with 3, or, mathematically speaking, $6 \div 2 = 3$. Now, let's take 6 beans and divide them into 3 equal groups. Mathematically speaking, it is $6 \div 3 = 2$.

Next, take 12 beans and divide these first into 2 equal groups, then into 3 equal groups, and then again into 4 equal groups. How many beans were in each group each time? *Ans:* 6, 4, and 3 beans; this is because $12 \div 6 = 2$, $12 \div 4 = 3$, $12 \div 3 = 4$.

If you remember the multiplication facts (for example $2 \times 2 = 4$, $3 \times 4 = 12$, $5 \times 3 = 15$, etc.), it will be easy for you to do division.

What is:

$2 \times 2 = 4$	$12 \div 3 = 4$	$10 \div 2 = 5$	$6 \times 3 = 18$
$4 \div 2 = 2$	$12 \div 4 = 3$	$10 \div 5 = 2$	$8 \div 2 = 4$
$3 \times 4 = 12$	$5 \times 2 = 10$	$6 \div 3 = 2$	$10 \div 2 = 5$

▶ *The rule: Remember that in multiplication (as in addition) the order is not important. That means that the answer for 3×2 will be the same as for 2×3, and 2×10 is equal to 10×2.*

Remember that this rule does not work for subtraction. That means the answer to $10 - 5$ will not be the same as $5 - 10$! In subtraction as in division (as you will learn), order is very very important and you can't switch the numbers around and get the same answer.

TIMES TABLE FOR 4 AND 3

$1 \times 4 = 4$	$4 \times 4 = 16$	$7 \times 4 = 28$	$10 \times 4 = 40$
$3 \times 1 = 3$	$3 \times 4 = 12$	$3 \times 7 = 21$	$3 \times 11 = 33$
$3 \times 4 = 12$	$4 \times 2 = 8$	$3 \times 1 = 3$	$3 \times 10 = 30$
$2 \times 4 = 8$	$5 \times 4 = 20$	$8 \times 4 = 32$	$11 \times 4 = 44$
$3 \times 2 = 6$	$3 \times 5 = 15$	$3 \times 8 = 24$	$3 \times 10 = 30$
$6 \times 4 = 24$	$4 \times 5 = 20$	$3 \times 4 = 12$	$3 \times 3 = 9$
$3 \times 4 = 12$	$6 \times 4 = 24$	$9 \times 4 = 36$	$12 \times 1 = 12$
$3 \times 3 = 9$	$3 \times 6 = 18$	$3 \times 9 = 27$	$2 \times 12 = 24$
$9 \times 4 = 36$	$4 \times 8 = 32$	$3 \times 7 = 21$	$6 \times 3 = 18$

EXERCISE I

$4 \times 3 = 12$	$4 \times 3 = 12$	$2 \times 7 = 14$	$11 \times 2 = 22$
$2 \times 4 = 8$	$16 + 4 = 20$	$9 \times 2 = 18$	$9 \times 2 = 18$
$12 - 3 = 9$	$4 \times 5 = 20$	$2 \times 9 = 18$	$6 \times 3 = 18$
$3 \times 3 = 9$	$8 \times 2 = 16$	$18 + 2 = 20$	$18 + 3 = 21$
$4 \times 4 = 16$	$2 \times 8 = 16$	$2 \times 10 = 20$	$3 \times 7 = 21$
$16 - 12 = 4$	$16 - 2 = 14$	$2 \times 11 = 22$	$6 \times 3 = 18$

5 × 3 = 15	3 × 8 = 24	8 × 3 = 24	3 × 9 = 27
15 + 3 = 18	3 × 7 = 21	9 × 3 = 27	27 - 3 = 24
18 + 3 = 21	3 × 6 = 18	3 × 6 = 18	30 - 9 = 21
21 + 3 = 24	21 + 3 = 24	3 × 7 = 21	30 - 12 = 28

WORD PROBLEMS

1. A turtle has four legs.
    a) How many legs does 1 turtle have? **Ans:** 4 legs.
    b) How many legs do 2 turtles have? **Ans:** 8 legs.
    c) How many legs do 3 turtles have? **Ans:** 12 legs.

2. How many legs do 4 chairs have? **Ans:** 16 legs.

3. There are four pencils in 1 box.
    a) How many pencils are in 3 boxes? **Ans:** 12 pencils.
    b) How many pencils are in 4 boxes? **Ans:** 16 pencils.
    c) How many pencils are in 5 boxes? **Ans:** 20 pencils.

4. There are 5 cents in one nickel.
    a) How many in 3 nickels? **Ans:** 15 cents.
    b) How many in 4 nickels? **Ans:** 20 cents.

5. If there are five fingers on one hand, how many fingers are on 4 hands? **Ans:** 20 fingers.

6. Mrs. Crumpet puts 4 spoons of sugar in her tea. How many spoons of sugar she will use for 5 cups? **Ans:** 16 spoons.

7. Mrs. Crumpet was having tea with her 3 friends. Her friends only like 2 teaspoons of sugar in their tea. So counting her 4 teaspoons, how many teaspoons of sugar will be used? **Ans:** 10.

8. If there are 3 eight-legged spiders in one corner,
    a) How many spiders are in 2 corners? **Ans:** 6 spiders.
    b) How many spiders are in 4 corners?
    Ans: 12 spiders. I hope you didn't count the legs.

9. Mr. Crumpet puts 8 pieces of salami on his sandwich. How many pieces of salami will he need for 2 sandwiches? **Ans:** 16 pieces of salami.

10. One week has 7 days. How many days are in three weeks?
    Ans: 21 days.

11. A football team made 3 touchdowns, each worth 6 points. How many points did the team score? *Ans:* 18 points.

12. A basketball player made 6 3-point shots. How many points did he score? *Ans:* 18 points.

13. If one play lasts 3 hours, how long will 7 plays last? *Ans:* 21 hours, but only 3 if they are all going on simultaneously. Simultaneously means at the same time.

14. For a soccer game, kids broke into 2 equal groups of 11 players each. How many kids played? *Ans:* 22 players.

15. Boxes of paper were delivered to the office. Each box has 8 reams of paper. How many reams are in 3 boxes? *Ans:* 24 reams. A ream of paper means 500 sheets of American size paper, about 2" high stack.

16. In a restaurant, a waiter put two forks on each placemat. How many forks did he put on 9 placemats? *Ans:* 18 forks.

17. For a fancy dinner, Mrs. Fancy put 3 forks on the left side of the plate. How many knives does she put on the right side? *Ans:* We don't know but it can be two or three.

18. A store sells vibrating toothbrushes for $3 each.
    a) How much do 8 toothbrushes cost? *Ans:* $24.
    b) How much do 9 toothbrushes cost? *Ans:* $27.

19. How many legs do 3 spiders have? *Ans:* 24 legs.

20. How many legs do 7 buffaloes have? *Ans:* 28 legs.

21. If a shark has 8 fins, how many fins do 3 sharks have?
    Ans: 24 fins.

22. If one crab has 2 claws, how many claws do 9 crabs have?
    Ans: 18 claws.

23. If a melon weighs 4 pounds, how much will 7 melons weigh?
    Ans: 28 lb.

24. If a mechanic can fix 4 cars in one day, how many cars can he fix in 6 days? *Ans:* 24 cars.

25. If a bicyclist can ride 9 miles in one hour, how many miles can he ride in 3 hours? *Ans:* 27 miles.

26. If each table seats 4 people, how many people can sit at 7 tables? *Ans:* 28 people.

27. Each box has 4 speakers. How many speakers are in 8 boxes? *Ans:* 32.

28. Each boat has 8 passengers.
    a) How many passengers are in 3 boats? *Ans:* 24 passengers.
    b) How many passengers are in 4 boats? *Ans:* 32 passengers.

29. A movie ticket costs $9. How much do 3 tickets cost? *Ans:* $27.

30. A cube has 6 sides. How many sides do 4 cubes have?
    Ans: 24 sides.

31. In a 4-story building each floor has 8 windows. How many windows are in the building? *Ans:* 32 windows.

32. An ice hockey game lasts 3 periods. How many periods are in 8 games? *Ans:* 24 periods.

33. There are 6 players on each ice hockey team. How many players are on 4 teams? *Ans:* 24 players.

34. An octagon has 8 sides.
    a) How many sides do 3 octagons have? *Ans:* 24 sides.
    b) How many sides do 4 octagons have? *Ans:* 32 sides.

35. If Jim likes to draw 3 pictures per page, how many pictures might he draw on 9 pages? *Ans:* 27 pictures.

36. A boy planted 8 rows of rose bushes, 4 bushes in each row. How many rose bushes did he plant? *Ans:* 32 rose bushes.

37. A teacher divided the class in groups of 3. If there are 10 groups, how many students are in this class? *Ans:* 30 students.

38. A piano student practices 3 hours every day.
    a) How many hours does she practice each week?
    Ans: 21 hours.
    b) How many hours does she practice in 8 days?
    Ans: 24 hours.
    c) How many hours does she practice in 9 days? *Ans:* 27 hours.

39. A mole digs 4 feet of underground tunnel in one hour.
    a) How many feet can it dig in 7 hours? *Ans:* 28 feet.
    b) How many feet can it dig in 8 hours? *Ans:* 32 feet.
    c) How many feet can it dig in 10 hours? *Ans:* 40 feet.

40. I can put 11 glasses on a tray. How many glasses can I put on 3 trays? *Ans:* 33 glasses.

41. If 1 can holds 4 large sardines, how many large sardines do 9 cans hold? *Ans:* 36 large sardines.

42. If the same can can hold 9 small sardines, how many sardines do 4 cans hold? *Ans:* 36 small sardines.

43. A secret door to a secret room has 5 locks. How many locks will be on 4 secret doors? *Ans:* 20 secret locks.

44. Four athletes compete in a race. How many athletes will compete in 7 races? *Ans:* 28 athletes.

45. If 8 college students can fit in one small car,
    a) How many can fit in 3 cars? *Ans:* 24 students.
    b) How many can fit in 4 cars? *Ans:* 32 students.

46. If 4 men can fit in a small truck,
    a) How many can fit in 7 small trucks? *Ans:* 28 men.
    b) How many can fit in 8 small trucks? *Ans:* 32 men.
    c) How many can fit in 9 small trucks? *Ans:* 36 men.

47. Margaret bought 4 shirts at $8 each. How much did she spend for all the shirts? *Ans:* $32.

48. Margaret also bought 4 matching pants at $9 each. How much did they cost? *Ans:* $36.

49. Conrad walked 7 miles each day.
    a) How many miles did he walk in 2 days? *Ans:* 14 miles.
    b) How many miles did he walk in 3 days? *Ans:* 21 miles.
    c) How many miles did he walk in 4 days? *Ans:* 28 miles.

50. There are 10 oz. of lemonade in a plastic cup.
    a) How many ounces of lemonade are in 2 cups? *Ans:* 20 oz.
    b) How many ounces of lemonade are in 3 cups? *Ans:* 30 oz.
    c) How many ounces of lemonade are in 4 cups? *Ans:* 40 oz.

51. There are 7 digits in a telephone number. How many digits are in 4 telephone numbers? *Ans:* 28 digits.

52. How many horseshoes do 8 horses need? *Ans:* 32 horseshoes.

53. How many horseshoes do 9 horses need? *Ans:* 36 horseshoes.

54. How many horseshoes do 6 lions need? *Ans:* None, lions don't wear horse shoes.

55. It takes 9 ants to carry one pine needle. How many ants does it take to carry 4 needles? *Ans:* 36 ants.

56. If one teenager sleeps for 11 hours, how many hours will it take for 4 of them? *Ans:* 44, but probably 11 hours, if they all go to sleep at the same time.

21

DIVISION BY 2 OR 3

Division is opposite of multiplication. If multiplication is combining sets (baskets) of numbers, then division is separating numbers into equal baskets.

Let's use the beans, as in lesson 17, to illustrate division. Take out 6 beans and divide them into 3 equal groups of 2. Say: "In mathematics, we say that we divided 6 beans in 2 groups and the answer is 3 beans in each group; or we divided 6 by 2, and that equals 3." Then divide 6 beans into 3 groups and show that the answer is 2 (the number of beans in each group).

Repeat this process with 8 beans (2 and 4), 10 beans (2 and 5), 12 beans (3 and 4, and 2 and 6), and also 24 beans (2 and 12, 6 and 4, 3 and 8).

Reminder: Division is opposite to multiplication. In multiplication we add together groups of the same numbers and count how many are in all the groups.

In division we separate a number into equal groups to find out the number in each group. Likewise, by knowing the number of each group, we can find the number of groups. There is another way to look at it. When we break a number into equal groups, and know the number that makes each group, we can tell the number of groups.

EXAMPLES:

If I have 12 greeting cards and divide them equally into 3 boxes, I can tell how many cards are in each box.

12 (cards) ÷ 3 (boxes or groups) = 4 (cards in each box or group)

On the other hand, if you know the number of cards in each box, you can find how many boxes were used.

12 (cards) ÷ 4 (cards in each box) = 3 (boxes).

Also remember that any number divided by 1 equals itself and any number divided by itself equals 1.

5 ÷ 1 = 5	1 ÷ 1 = 1	48 ÷ 1 = 48
5 ÷ 5 = 1	21 ÷ 21 = 1	48 ÷ 48 = 1

The only exception to this rule is 0, because you should never never divide by zero.

EXERCISE I

1 × 2 = 2	6 × 2 = 12	2 ÷ 2 =1	12 ÷ 2 = 6
2 × 2 = 4	7 × 2 = 14	4 ÷ 2 = 2	14 ÷ 2 = 7
3 × 2 = 6	8 × 2 = 16	6 ÷ 2 = 3	16 ÷ 2 = 8
4 × 2 = 8	9 × 2 = 18	8 ÷ 2 = 4	18 ÷ 2 = 9
5 × 2 = 10	10 × 2 = 20	10 ÷ 2 = 5	20 ÷ 2 = 10

Sometimes, instead of saying "divide by two" we say "find one half". If we ask how much is one half of something, we mean how much will be in each part if divide it by 2.

For example:

- What is one half of 6?
  Ans: 3, because if divide 6 in two parts, each part will equal 3.

- What is one half of 8? *Ans:* 4.

- What is one half of 10? *Ans:* 5.

- What is one half of 2? *Ans:* 1.

- What is one half of 14? *Ans:* 7.

Reminder: In multiplication the order is not important, 2×5 is the same as 5×2.

In division (as in subtraction) we can't switch positions of the numbers. That means $10 \div 2$ in not the same as $2 \div 10$. Not at all!

EXERCISE II

- How many 2's are in 2? *Ans:* 1.
- How many 2's are in 8? *Ans:* 4.
- How many 2's are in 6? *Ans:* 3.
- How many 2's are in 4? *Ans:* 2.
- How many 2's are in 12? *Ans:* 6.
- How many 2's are in 20? *Ans:* 10.
- How many 2's are in 16? *Ans:* 8.
- How many 2's are in 14? *Ans:* 7.
- How many 2's are in 10? *Ans:* 5.
- How many 2's are in 18? *Ans:* 9.
- How many 3's are in 3? *Ans:* 1.
- How many 3's are in 9? *Ans:* 3.
- How many 3's are in 21? *Ans:* 7.
- How many 3's are in 15? *Ans:* 5.
- How many 3's are in 12? *Ans:* 4.
- How many 3's are in 18? *Ans:* 6.
- How many 3's are in 6? *Ans:* 2.
- How many 3's are in 24? *Ans:* 8.
- How many 3's are in 30? *Ans:* 10.
- How many 3's are in 27? *Ans:* 9.

EXERCISE III

$10 \div 2 = 5$	$2 \times 8 = 16$	$18 \div 2 = 9$	$20 \div 2 = 10$
$12 \div 2 = 6$	$14 \div 2 = 7$	$2 \times 4 = 8$	$8 \div 2 = 4$

2 × 6 = 12	30 ÷ 3 = 10	3 × 6 = 18	21 ÷ 3 = 7
16 ÷ 2 = 8	6 ÷ 3 = 2	24 ÷ 3 = 8	9 × 3 = 27
2 × 7 = 14	12 ÷ 3 = 4	2 × 3 = 6	9 ÷ 3 = 3
5 × 2 = 10	3 × 6 = 18	3 × 4 = 12	12 ÷ 3 = 4
9 × 2 = 18	3 × 3 = 9	18 ÷ 3 = 6	12 ÷ 4 = 3
2 × 3 = 6	21 ÷ 3 = 7	3 × 8 = 24	18 ÷ 3 = 6
7 × 2 = 14	5 × 3 = 15	3 × 5 = 15	18 ÷ 9 = 2

WORD PROBLEMS

1. I took 6 cards and divided them into two equal stacks. How many cards are in each stack? *Ans:* 3 cards.

2. During the game two teams scored 8 goals. The game was a tie. How many goals did each team score? *Ans:* 4 goals.

3. A tailor took 10 feet of fabric and cut it into 2 equal pieces. How long is each piece? *Ans:* 5 feet.

4. Leslie had 8 puppies and gave half of them to her friends. How many puppies did she keep? *Ans:* 4 puppies.

5. An athlete went on a 10-mile run and took a short rest exactly half way. How many miles did she run before stopping? *Ans:* 5 miles.

6. Nisha's walk to school and back to home (the round-trip) takes 12 minutes. How long is her one-way walk? *Ans:* 6 minutes.

7. To play a hockey game, 12 children split into 2 teams. How many children are in each team? *Ans:* 6 kids.

8. If a pound of coffee cost $10, how much would a half a pound cost? In other words, if we divide a pound of coffee into two equal parts, how much does each part cost? *Ans:* $5.

9. There are 10 fingers on both hands, how many fingers are on one hand? *Ans:* 5.
    Was that easy? I thought so.

10. There are 20 teeth in the mouth, half on top and the other half on the bottom.
a) How many teeth are on top? *Ans:* 10 teeth.
b) How many on the bottom? *Ans:* 10 teeth.
That was easy, wasn't it?

11. In the zoo I counted 12 penguin legs. How many penguins were in the zoo? *Ans:* 6 penguins, because each penguin has 2 legs.

12. Twenty-two penguins are taking care of their eggs by carrying them on both feet. How many eggs are they taking care of? *Ans:* 22 eggs.

13. If two toothpaste tubes weigh 14 oz., how much does each tube weigh? *Ans:* 7 oz.

14. Newborn baby twins together weigh 14 pounds. How much does each twin weigh?
Ans: 7 pounds each, but most of the time twins are born with different weight; and it is not unusual for one twin to weigh 8 lb. and the other 6 lb. which together would also make 14 lb.

15. A pharmacist has to divide the medicine equally into two bottles. If there are 16 pills, how many will go into each bottle? *Ans:* 8 pills.

16. A king wants to divide 9 castles equally among his 3 sons. How many castles will each son get? *Ans:* 3 castles.

17. The king has 16 farms and wants to divide them equally between his two daughters. How many farms will each daughter receive? *Ans:* 8 farms.

18. There are 16 black and white pawns on a chess board. How many pawns of each color are on the board? *Ans:* 8 pawns.

19. An explorer divided 9 cans of spinach equally for 3 days of mountain travel. How many cans of spinach did he eat each day? *Ans:* 3 cans.

20. Three friends shared 12 pears. How many pears did each friend get? *Ans:* 4 pears.

21. Lisa threw 9 coins into 3 fountains, the same number of coins each time. How many coins went into each fountain? *Ans:* 3 coins.

22. After an accident, 12 people were divided equally to go to 2 emergency rooms. How many went to ER? *Ans:* 6 people. ER stands for Emergency Room.

23. Abbey invited an equal number of boys and girls to her party. All 18 invitees came. How many girls came to the party? *Ans:* 9 girls. Invitees means people who were invited.

24. Karim bought 12 bagels and on the way home ate one half. How many bagels did he bring home? *Ans:* 6 bagels.

25. On an 18 wheel truck, how many wheels are on each side? *Ans:* 9 wheels.

26. To make 3 cakes, a chef divided 12 eggs equally. How many eggs went into each cake? *Ans:* 4 eggs.

27. Latisha has $16, all in $2 bills. How many $2 bills does she have? *Ans:* 8 bills. $2 bills are hard to find. Many people collect them.

28. If 3 trucks have to deliver 15 bags of mail, how many bags does each truck carry? *Ans:* 5 bags.

29. If 15 acres of land are divided into 3 lots, how many acres are in each lot? *Ans:* 5 acres. If you don't know what acres are - don't worry as long as you have got the right answer. A"lot" in the U.S.A. mean a building area where just one house or building can be built.

30. An Olympic runner ran 18 miles in 2 hours. How many miles did the runner run in 1 hour? *Ans:* 9 miles.

31. A runner ran 26-mile marathon in 2 hours. How many miles did he run in one hour? *Ans:* 13 miles. Did you know that the world record for running a 26 mile marathon is two and a half hours?

32. A painter painted 3 classrooms in 15 hours. How long did it take to paint each room? *Ans:* 5 hours.

33. A locksmith installed 14 locks in one day, 2 locks on each door. How many doors got the locks? *Ans:* 7 doors, because he put 14 locks altogether. He divided all locks into groups of 2. He came out with seven groups, $14 \div 2 = 7$.

34. James built 18 feet of a wall in 3 days. How many feet of the wall did he build each day? *Ans:* 6 feet.

35. Bob and his 2 friends went out for lunch and the bill came to $18. They split it three ways. How much did each pay? *Ans:* $6. The bill is $18 and it was divided equally between 3 friends, $18 \div 3 = 6$.

36. A school purchased 3 new desks and paid $24 for all three. How much did each desk cost? *Ans:* $8.

37. If there are 20 gloves in a store, how many pairs of gloves does it make? *Ans:* 10 pairs, because 2 gloves make one pair. If we take 20 gloves and divide them into groups of 2, then we'll have 10 pairs.

38. A carpenter took a 21-foot board and cut it into 3 equal pieces. How long was each piece? *Ans:* 7 feet.

39. Three fishermen caught 24 pounds of fish and divided it among themselves. How many pounds did each fisherman get? *Ans:* 8 lb.

40. A restaurant paid $18 for 3 bags of tomatoes. How much did each bag cost? *Ans:* $6.

41. The restaurant paid $18 for 2 boxes of mushrooms. What was the price of each box? *Ans:* $9.

42. A three-headed dragon altogether had 30 teeth. How many teeth did each head have? *Ans:* 10 teeth.

43. An interior designer bought 24 pillows for 3 sofas. How many pillows did she buy for each sofa? *Ans:* 8 pillows.

44. If each page of a photo album can hold 3 photographs, how many pages will you need to fit 21 pictures? *Ans:* 7 pages. *Solution:* There are 21 photograph that we divide into groups of 3, $21 \div 3 = 7$. It will take 7 pages to place all the photographs, 3 pictures per page.

45. There are 16 fruits in a bowl; half are oranges and the rest are apples. How many apples are in the bowl? *Ans:* 8 apples.

46. A group of 18 workers split into 3 equal teams to build a fence. How many workers were in each team? *Ans:* 6 workers.

47. How many candles can you buy for $15 if each candle costs $3? *Ans:* 5 candles.

48. How many toothbrushes can you buy for $14 if each brush costs $2? *Ans:* 7 toothbrushes.

49. How many tennis balls will fit into one paper bag if we divide 20 balls equally into 2 bags? *Ans:* 10 tennis balls.

50. If 27 kids divide into 3 equal number teams, how many will be on each team? *Ans:* 9 children.

51. Three equal size backpacks together weigh 21 pounds. How much does each bag weigh? *Ans:* 7 lb.

52. The grandpa bought 2 pairs of glasses for $20. How much was one pair? *Ans:* $10. Don't be confused with word "pair" which means only one item.

53. Jackie's grandma bought 3 pairs of sun glasses and paid $27. What was the price of one pair of glasses? *Ans:* $9.

22

MULTIPLICATION AND DIVISION BY 2, 3, OR 4

Mixing multiplication and division problems is teaching the child to listen carefully to the problems rather than jumping to solve them. Prepare to read the problem two or more times to get "all the facts" straight.

Some of these problems can be a bit confusing, for example, compare these two:

1. If a baker can make 6 cakes in one day, how many cakes can 3 bakers make? Answer to this problem is 18 cakes (we multiply).

2. If it takes one kid 4 hours to build a sandcastle, how long will it take 2 kids? The answer to this problem is 2 hours (we divide).

One problem required multiplication and the other division. Offer your help to solve them and don't be discouraged if it takes time for your child to grasp these concepts.

EXERCISE I

3 + 6 = 9	21 + 6 = 27	4 + 8 = 12	32 - 8 = 24
9 + 6 = 15	9 + 9 = 18	8 + 8 = 16	32 - 16 = 16
6 + 6 = 12	9 + 9 + 9 = 27	8 + 8 + 8 = 24	4 + 32 = 36
9 + 9 = 18	9 + 18 = 27	12 + 8 = 20	8 + 24 = 32
18 - 12 = 6	30 - 3 = 27	16 + 4 = 20	36 - 24 = 12
18 + 12 = 30	30 - 6 = 24	16 + 8 = 24	12 + 12 = 24
21 - 12 = 9	30 - 9 = 21	20 - 8 = 12	36 - 18 = 18
12 + 9 = 21	27 - 12 = 15	24 - 12 = 12	40 - 12 = 28
27 - 3 = 24	30 - 12 = 18	36 - 9 = 27	40 - 24 = 16
27 - 9 = 18	15 + 15 = 30	28 - 16 = 12	32 - 16 = 16

EXERCISE II

1. How many 3's are in 3? *Ans:* 1
2. How many 3's are in 9? *Ans:* 3
3. How many 3's are in 6? *Ans:* 2
4. How many 3's are in 18? *Ans:* 6
5. How many 3's are in 15? *Ans:* 5
6. How many 3's are in 21? *Ans:* 7
7. How many 3's are in 24? *Ans:* 8
8. How many 3's are in 30? *Ans:* 10
9. How many 3's are in 27? *Ans:* 9
10. How many 3's are in 12? *Ans:* 4
11. How many 3's are in 33? *Ans:* 11

EXERCISE III

1 × 4 = 4	7 × 4 = 28	12 ÷ 4 = 3	36 ÷ 4 = 9
2 × 4 = 8	8 × 4 = 32	16 ÷ 4 = 4	40 ÷ 4 = 10
3 × 4 = 12	9 × 4 = 36	20 ÷ 4 = 5	3 × 4 = 12
4 × 4 = 16	10 × 4 = 40	24 ÷ 4 = 6	32 ÷ 8 = 4
5 × 4 = 20	4 ÷ 4 =1	28 ÷ 4 = 7	36 ÷ 6 = 6
6 × 4 = 24	8 ÷ 4 = 2	32 ÷ 4 = 8	40 ÷ 8 = 5

EXERCISE IV

6 × 2 = 12	20 ÷ 2 = 10	24 ÷ 2 = 12	4 × 6 = 24
12 ÷ 4 = 3	3 × 7 = 21	9 × 2 = 18	24 ÷ 8 = 3
12 ÷ 2 = 6	3 × 8 = 24	3 × 6 = 18	8 × 4 = 32
4 × 4 = 16	16 ÷ 4 = 4	16 ÷ 4 = 4	4 × 7 = 28
16 ÷ 2 = 8	16 ÷ 2 = 8	18 ÷ 3 = 6	3 × 9 = 27
6 × 3 = 18	4 × 6 = 24	7 × 3 = 21	20 ÷ 2 = 10
18 ÷ 3 = 6	3 × 8 = 24	7 × 4 = 28	6 × 4 = 24
18 ÷ 2 = 9	24 ÷ 4 = 6	3 × 7 = 21	7 × 5 = 35
5 × 4 = 20	24 ÷ 3 = 8	4 × 3 = 12	

EXERCISE V

- 2 × 3 = 6; What other numbers can you multiply together to make 6? *Ans:* 6 × 1 = 6.
- 2 × 4 = 8; What other numbers can you multiply together to make 8? *Ans:* 8 × 1 = 8.
- 4 × 3 = 12; What other numbers can you multiply together to make 12? *Ans:* 6 × 2 = 12 and 12 × 1 = 12.
- 4 × 4 = 16; What other numbers can you multiply together to make 16? *Ans:* 8 × 2 = 16 and 16 × 1 = 16.
- 4 × 6 = 24; What other numbers can you multiply together to make 24? *Ans:* 3 × 8 = 24 and 12 × 2 = 24.
- 4 × 5 = 20; What other numbers can you multiply together to make 20? *Ans:* 2 × 10 = 20 , 20 × 1.
- 15 × 2= 30; What other numbers can you multiply together to make 30? *Ans:* 2 × 15 = 30, 30 × 1, 6 × 5.

WORD PROBLEMS

1. If 1 hockey puck costs $2, how much will 2 hockey pucks cost? *Ans:* $4.

2. At $2 each, how many hockey pucks can I buy for $6?
    Ans: 3 hockey pucks.
    Solution: I have $6 that I need to divide in my mind into groups
    of $2 (each group is the price of one puck). There are $6 ÷ $2 =
    3 (or 3 groups holding the price of one puck each).
    a) How many hockey pucks can you buy for $8?
    Ans: 4 hockey pucks.
    b) How many hockey pucks can you buy for $10?
    Ans: 5 hockey pucks.
    c) How many hockey pucks can you buy for $18?
    Ans: 9 hockey pucks.

3. Eight pilgrims carried 2 loaves of bread each. How many loaves
    of bread did they carry? *Ans:* 16 loaves.
    Solution: Each pilgrim has 2 loaves and there are 8 pilgrims.
    Then, 2 (loaves) × 8 (pilgrims) = 16 (loaves, not pilgrims, of
    course).

4. Wayne made 6 cups of tea and put 3 lumps of sugar in each
    cup. How many lumps of sugar did he use? *Ans:* 18 lumps.

5. How many 3 inch pieces can I cut from a 12 inch long ribbon?
    Ans: 4 pieces.

6. How many 4 inch pieces can I cut from a 12 inch long ribbon?
    Ans: 3 pieces.

7. How many 6 inch pieces can I cut from a 12 inch long ribbon?
    Ans: 2 pieces.

8. Ann has 6 books, Bob has 2 times as many. Bob has how many
    books? *Ans:* 12 books.

9. How many apples can I put in each bag if I have 20 apples and
    4 bags? *Ans:* 5 apples in each bag, because there are 20 apples
    that I divided into 4 groups, 20 ÷ 4 = 5.

10. Archie earns $6 per hour.
    a) How much will he make in 3 hours? *Ans:* $18.
    b) How much will he make in 4 hours? *Ans:* $24.

11. Jerry has 8 toys. Tom has 2 times as many toys as Jerry. How
    many toys does Tom have? *Ans:* 16 toys.

12. A rule says "measure twice (two times), cut once."
    a) How many times should you measure to make 2 cuts?
    Ans: 4 times.
    b) How many times should you measure to make 5 cuts?
    Ans: 10 times.
    Solution: If I cut 1 time, I should measure 2 times. Right? If I cut 2 times, I should measure 4 times, and so on. If I cut 5 times, I should measure 10 times, still 2 times of measure for each cut.
    c) How many times should you measure to make 8 cuts?
    Ans: 16 times.
    d) How many times should you measure to make 9 cuts?
    Ans: 18 times.
    e) How many times should you measure to make 10 cuts?
    Ans: 20 times.

13. Three Marines divided a 24-hour guard duty into 3 shifts. How much time did each Marine get for guard duty? *Ans:* 8 hours.

14. Mom left the house at 8 A.M., I left 2 hours earlier. What time did I leave? *Ans:* 6 A.M., because 8 A.M. - 2 hours = 6 A.M.

15. Mom left the house at 8 A.M., dad left 2 hours later. What time did he leave? *Ans:* 10 A.M.

16. If 4 kids have 2 bunnies each, how many bunnies do they have together? *Ans:* 8 bunnies.

17. How many ears do 4 horses have? *Ans:* 8 ears.

18. How many legs do 4 horses have? *Ans:* 16 legs.

19. Lucia poured 18 ounces of juice into 3 glasses. How many ounces of juice is in each glass? *Ans:* 6 oz.

20. Three workers were given $24 which they divided equally. How much did each receive? *Ans:* $8.

21. Jeremy studied for a test 3 hours every day. How many hours he spent studying in 6 days? *Ans:* 18 hours.

22. Four friends ate 8 slices of pizza. How many pieces did each friend eat? *Ans:* 2 pieces.

23. Asha needs to work 10 hours on a project. She broke the work down into 2 days. How much time will she work on her project each day? *Ans:* 5 hours a day.

24. Habib borrowed 16 spoons and returned 8. How many spoons does he still have to return? *Ans:* 8 spoons.

25. If a squirrel gathers 4 nuts a day, how long will it take him to gather 12 nuts? *Ans:* 3 days.

26. For exercise, my mom ran 4 miles. I can run 3 times as much. How many miles do I think I can run? *Ans:* 12 miles.

27. We need to divide 16 treats between 2 puppies. How many treats will each puppy get? *Ans:* 8 treats.

28. Now, we need to divide 16 treats among 4 puppies. How many treats will each get? *Ans:* 4 treats.

29. I used 3 colored pencils, but my sister used 6 times as many. How many colored pencils did she use? *Ans:* 18.

30. A man walked 9 miles in 3 hours. How many miles did he walk each hour? *Ans:* 3 miles.

31. Nick bought 4 chocolate bars for his 4 friends. How many chocolate bars did each friend get? *Ans:* 1 chocolate bar.

32. One book has 9 chapters. How many chapters do 3 books have? *Ans:* 27 chapters, if they are the same book.

33. If there are 4 sardines in one can, how many sardines are in 8 cans? *Ans:* 32 sardines.

34. If one sardine has 2 eyes, how many eyes do 8 sardines have? *Ans:* 16 eyes.

35. One barber can shave 4 beards in an hour.
a) How many beards can 4 barbers shave? *Ans:* 16 beards.
b) How many beards can 1 barber shave in 4 hours?
Ans: 16 beards.

36. One locksmith can install 3 locks in 1 hour.
    a) How many locks can 7 locksmiths install? *Ans:* 21 locks.
    b) How many locksmiths will it take to install 18 locks in 1 hour? *Ans:* 6 locksmiths.
    c) How many locksmiths will it take to install 27 locks in 1 hour? *Ans:* 9 locksmiths.
    d) How many locks can 1 locksmith install in 5 hours? *Ans:* 15 locks.
    e) How many locks can 1 locksmith install in 8 hours? *Ans:* 24 locks.

37. A watchmaker can repair 4 grandfather clocks in 1 week.
    a) How many clocks can he repair in 5 weeks? *Ans:* 20 clocks.
    b) How many clocks can he repair in 7 weeks? *Ans:* 28 clocks.
    c) How many clocks can he repair in 6 weeks? *Ans:* 24 clocks.

38. It takes 2 minutes for a receptionist to take a message. How much time will it take her to take 5 messages? *Ans:* 10 minutes.

39. If a tricycle has 3 wheels, how many wheels do 5 tricycles have? *Ans:* 15 wheels.

40. If a car has 4 wheels, how many wheels do 3 cars have? *Ans:* 12 wheels.

41. If 2 men together earned $20, how much money will 3 men earn? *Ans:* $30.
    Solution: If 2 men earned $20, then 1 man earned $20 ÷ 2 = $10. If one man earned $10, then 3 men earned $10 × 3 = $30.

42. If there are 4 pens in 2 boxes, how many pens are in 5 boxes? *Ans:* 10 pens.

43. If two handcarts can be rented for $8, then how much will it cost to rent three handcarts? *Ans:* $12.
    Solution: One handcart cost $8 ÷ 2 = $4.
    So 3 will cost $4 × 3 = $12.

44. One classroom has 6 tables, how many tables would be in 2 classrooms? *Ans:* 12 tables.

45. If each cow has 4 legs, how many legs do 4 cows have?
Ans: 16 legs.
a) How many legs do 5 cows have? *Ans:* 20 legs.
b) How many legs do 6 cows have? *Ans:* 24 legs.
c) How many legs do 7 cows have? *Ans:* 28 legs.

46. How many toy cars can you build using 16 wheels if each car needs 4 wheels? *Ans:* 4 cars.
a) How many toy cars can you build with 20 wheels?
Ans: 5 cars.
b) How many toy cars can you build with 24 wheels?
Ans: 6 cars.
c) How many toy cars can you build with 32 wheels?
Ans: 8 cars.

47. If each toy car has 4 wheels, how many wheels are on 9 cars?
Ans: 36 wheels.

48. If each room has 4 chairs, how many chairs are in 3 rooms?
Ans: 12 chairs.

49. Grandma's bracelet has 8 pearls, her necklace 4 times as many. How many pearls are in grandma's necklace?
Ans: 32 pearls.

50. Coffee from a pitcher was poured into 4 mugs. How much coffee was in each mug if the pitcher held 32 ounces?
Ans: 8 ounces.

51. A blues singer recorded 36 songs and divided them equally among 4 CDs. How many songs were on each CD?
Ans: 9 songs.

52. A tailor bought 28 buttons for 4 shirts. How many buttons went to each shirt? *Ans:* 7 buttons.

53. A convention center has 6 rest-rooms on each floor. How many rest-rooms are on all 4 floors? *Ans:* 24 rest-rooms.

54. There are 5 bird's nests on an old oak tree. Each nest has 4 baby birds in it. How many baby birds are on the old tree?
Ans: 20 baby birds.

55. A mother bird brought 24 worms and gave each chick equal number of worms. How many worms did each of her 4 chicks get? *Ans:* 6 worms.

56. In a Make-a-Bear class 8 children made 4 bears each. How many bears did the whole class make? *Ans:* 32 bears.

57. Dr. Sweet always puts 4 teaspoons of sugar in his cup of tea.
    a) If he drank 7 cups of tea today, how many teaspoons of sugar did he use? *Ans:* 28 teaspoons.
    b) If he drank 8 cups of tea today, how many teaspoons of sugar did he use? *Ans:* 32 teaspoons.
    c) If he drank 9 cups of tea today, how many teaspoons of sugar did he use? *Ans:* 36 teaspoons.

58. At a beach, 32 children broke into teams of 4 to build sand castles. How many sand castles were built that day? *Ans:* 8.

59. While playing with his sister, Dennis broke 6 vases. Each vase split into 4 pieces. How many broken pieces did mom find? *Ans:* 24 pieces.

60. If one owl hoots 7 times, how many times will 4 owls hoot? *Ans:* 28 times.

23

MIXED OPERATIONS WITH 2, 3, OR 4

EXERCISE I

- How many 3's are in 6? *Ans:* 2
- How many 4's are in 4? *Ans:* 1
- How many 3's are in 12? *Ans:* 4
- How many 4's are in 12? *Ans:* 3
- How many 4's are in 16? *Ans:* 4
- How many 4's are in 24? *Ans:* 6
- How many 3's are in 24? *Ans:* 8
- How many 4's are in 8? *Ans:* 2
- How many 4's are in 24? *Ans:* 6
- How many 4's are in 32? *Ans:* 8
- How many 4's are in 16? *Ans:* 4
- How many 4's are in 20? *Ans:* 5
- How many 4's are in 28? *Ans:* 7
- How many 4's are in 40? *Ans:* 10
- How many 4's are in 44? *Ans:* 11 (there ten 4's in 40, and also another 4, hence 11) Count down from 30 by 10's.

EXERCISE II

$4 \times 3 = 12$	$24 \div 3 = 8$	$30 \div 3 = 10$	$40 \div 4 = 10$
$4 \times 6 = 24$	$24 \div 4 = 6$	$21 \div 3 = 7$	$14 \div 2 = 7$
$4 \times 4 = 16$	$28 \div 4 = 7$	$18 \div 3 = 6$	$18 \div 2 = 9$
$5 \times 4 = 20$	$27 \div 3 = 9$	$16 \div 2 = 8$	$9 \times 2 = 18$
$4 \div 4 = 1$	$7 \times 4 = 28$	$9 \times 4 = 36$	$5 \times 3 = 15$
$16 \div 4 = 4$	$2 \times 9 = 18$	$32 \div 4 = 8$	$6 \times 5 = 30$
$4 \times 0 = 0$	$6 \times 3 = 18$	$28 \div 4 = 7$	$40 \times 2 = 80$

EXERCISE III

- Count up from 31 to 55 by adding 4 (35, 39, 43, 47, 51, 55).
- Count up from 10 to 50 by adding 5 (10, 15, 20, 25, 30, 35, etc.).
- Count up from 22 to 47 by adding 5 (27, 32, 37, 42, 47).
- Count down from 76 to 48 by subtracting 4 (72, 68, 64, 60, 56, 52, 48).
- Count up from 24 to 66 by adding 6 (30, 36, 42, 48, 54, 60, 66).
- Count down from 99 to 63 by subtracting 6 (93, 87, 81, 75, 69, 63).

WORD PROBLEMS

1. If one couple invites 3 other couples for dinner, how many people will be at the dinner? *Ans:* 8 people.
    Solution: One couple is 2 persons, 3 couples is $2 \times 3 = 6$ persons. Then $2 + 6 = 8$ people. There is another way to solve this problem. We can count couples first: $1 + 3 = 4$ couples. Since each couple is 2 people, then 2 (people in each couple) × 4 (couples) = 8 people.

2. Jim found 4 gold nuggets. Josh found 5 and Judd found 3. They put all their gold nuggets together and then divided them equally. How many nuggets did each get?
    Ans: 4 nuggets for each.
    Solution: $4 + 5 + 3 = 12$ (acorns together).
    Then $12 \div 3 = 4$ (acorns for each boy).

3. Janice borrowed $13 from her mom and $5 from her brother to buy 2 sets of guitar strings. How much did each set of strings cost? **Ans:** $9.
    Solution: $13 + $5 = $18. Janice borrowed $18. If she bought 2 sets, then $18 ÷ 2 = $9 for each set of strings.

4. One morning 7 boys and 11 girls got together, split evenly into 2 teams and played soccer. How many children were on each team? **Ans:** 9 children (7 + 11 = 18, then 18 ÷ 2 = 9)

5. Liz took 4 biscuits and put them together with 10 other biscuits. Then she divided them into 2 plates. How many biscuits are on each plate? **Ans:** 7 biscuits.
    Solution: 4 + 10 = 14 (biscuits). Then 14 ÷ 2 = 7 (biscuits).

6. A coach invited 8 girls and 12 boys and divided them into 2 teams. How many children are in each team? **Ans:** 10 children.

7. Mr. Coo ordered 16 small and 4 large windows. He divided them evenly into 4 sets for each floor of the house. How many windows are on each floor? **Ans:** 5 windows.
    Solution: 16 + 4 = 20 (windows, large and small). 20 ÷ 4 (floors) = 5 windows per floor.

8. Out of 20 apples in a box, Naomi threw away 8 rotten apples and then divided the rest into 2 equal piles. How many apples are in each pile? **Ans:** 6 apples.
    Solution: 20 (apples) - 8 (rotten apples) = 12 apples. Then, 12 ÷ 2 = 6 apples in each pile.

9. There were 23 tourists in a hotel; 5 got sick and decided to rest. The rest split into 2 groups to go on two different outings? How many tourists were in each group? **Ans:** 9 tourists
    Solution: 23 (tourists) - 5 (sick) = 18 (healthy tourists). Then 18 ÷ 2 (groups) = 9 (tourists in each group) .

10. Billy found 25 candies in a bag. He quickly ate 9 and shared the rest with his friend. How many candies did his friend get? **Ans:** 8 candies.
    Solution: 25 (candies) - 9 (candies that Billy ate) = 16 (candies still in the bag). Then, 16 (candies) ÷ 2 = 8 (candies for the friend).

11. A bird caught 17 caterpillars. She ate 5 and divided the rest among her 3 baby birds. How many caterpillars did each baby get? *Ans:* 4 caterpillars.
    Solution: 17 - 5 = 12 (caterpillars), then 12 ÷ 3 = 4 (caterpillars to each baby bird).

12. Two puppies found 18 shoes. They divided the shoes in 2 equal piles and chewed 4 shoes from each pile. How many shoes were spared (not chewed)? *Ans:* 10 shoes.
    Solution: 18 (shoes) ÷ 2 = 9 (shoes in each pile). Then, 9 (shoes in a pile) - 4 (chewed shoes) = 5 (shoes, not chewed). But remember there are 2 piles, then 5 × 2 = 10 shoes.

13. How many legs do 3 birds and 2 rabbits have? *Ans:* 14 legs.
    Solution: 2 (legs per bird) × 3 (bird) = 6 leg (all the birds). 4 (legs per rabbit) × 2 (rabbits) = 8 legs (all the rabbits). 6 (bird legs) + 8 (rabbit legs) = 14 legs.

14. A store received a shipment of 2 boxes with new dresses, 6 dresses in each box. Then it took all the dresses out and divided them for display in 3 store window. How many dresses went into each window? *Ans:* 4 dresses.
    Solution: The store received 6 (dresses per box) × 2 (boxes) = 12 dresses in both boxes. Then it divided all 12 dresses into 3 groups for display, 12 ÷ 3 = 4. Therefore, there are 4 dresses in each window.

15. We saw 4 groups of birds, 4 birds in each group. They flew away, then came back and sat together in 2 groups. How many birds are in each group? *Ans:* 8 birds.
    Solution: 4 (birds) × 4 groups = 16 birds, then 16 (birds) ÷ 2 (new groups) = 8 (birds in each group).

16. Olga bought 4 chickens for $3 each and 3 ducks for $4 each. How much money did she spend? *Ans:* $24.
    Solution: She bought 4 chicken × $3 = $12. She also bought 3 ducks × $4 = $12. Together, $12 + $12 = $24.

17. A man had $44 out of which he spent $16 for gas. The rest of the money he divided equally to buy lunches over the next 4 days. How much money is he prepared to spend for each lunch? *Ans:* $7.
    Solution: $44 - $16 = $28. Then $28 ÷ 4 = $7.

18. There were 37 magazines on the shelf. Daisy threw away 15 and the rest she divided into 2 equal piles. How many magazines are in each pile? *Ans:* 11 magazines.

19. Ali found 7 golf balls, Ben found 8, and Jack found 6. They put all their golf balls together and then split all of them evenly among themselves. How many golf balls did each person receive? *Ans:* 7 golf balls.

20. Take a 31-foot wire, cut off 13 feet. Then cut the rest into 3 equal parts. How long is each part? *Ans:* 6 feet.

21. February has 28 days divided into 4 weeks. How many days does each week in February have? *Ans:* 7 days.

22. There are 12 months in a year divided into 4 seasons. How many months are in each season? *Ans:* 3 months ($12 \div 4 = 3$).

23. There were 33 people in line for taxicabs. Soon, 6 taxicabs came and each took 5 passengers. How many people are still waiting for cabs? *Ans:* 3 people.
    Solution: First, let's find how many people left in taxi cabs, 5 (people) × 6 (cabs) = 30 (people). Then, 33 (people waiting) - 30 (people who left) = 3 (people).

24. There were 40 folding chairs at the community pool. In one month, 14 chairs broke, and 6 were stolen. The remaining chairs were divided evenly into 4 stacks. How many chairs were in each stack? *Ans:* 5 chairs.
    Solution: Let's first find out how many chairs were lost because they were either broken or stolen, 14 + 6 = 20 (chairs). Then, we need to know the number of chairs left, 40 - 20 = 20 (chairs). These 20 chairs were divided into 4 stacks, 20 ÷ 4 = 5 (chairs in each stack).

25. A restaurant has 7 tables for 4 people each. There are 7 people working in the restaurant every night. How many people are in the restaurant when it's full? *Ans:* 35 people.
    Solution: When the restaurant is full, there are 4 (people at each table) × 7 (tables) = 28 (people). There are also 7 people in the staff, altogether 28 + 7 = 35 people.

26. Every morning for 4 days Anna-Maria made 3 sandwiches. On the 5th day she made 5. How many sandwiches did she make altogether? *Ans:* 17 sandwiches.
Solution: 3 (sandwiches) × 4 (days) = 12 (sandwiches). Then, 12 (sandwiches) + 5 sandwiches (the 5th day) = 17 (sandwiches).

27. In the emergency room, Dr. Singh saw 6 patients every hour for the first 3 hours. During the 4th hour, he saw 10 patients. How many patients did he see? *Ans:* 28 patients (6 × 3 = 18, then 18 + 10 = 28).

28. There were 16 pieces for each player on the chess board. After 14 moves one player lost 4 chess pieces and the other lost 5. How many chess pieces are on the board now? *Ans:* 23 pieces.
Solution: If there are 16 pieces for each player on each side of the board, then there are 16 + 16 = 32 pieces on the board. During the game (the number of moves is there to confuse you) the players lost 4 + 5 = 9 pieces. Then, the number of pieces left on board is 32 - 9 = 23 (pieces).

29. An army general earned 47 medals. He put 29 medals on his parade uniform and the rest he stored in 3 boxes. How many medals are in each box? *Ans:* 6 medals.
Solution: The general has 47 medals. After he put 29 medals on uniform, there were 47 - 29 = 18 (medals left to be put into boxes). Then 18 medals divided into 3 boxes are 18 ÷ 3 = 6 (medals in each box).

30. Bo washed 36 utensils. She put 12 knives in one drawer. The rest, spoons and forks, she divided equally into 4 groups. How many utensils were in each group? *Ans:* 6 utensils.
Solution: We will call all knives, spoons, and forks utensils. Bo washed 36 of them and put 12 (knives) in a drawer, so she had 36 - 12 = 24 (utensils left). Then, she divided 24 utensils in 4 groups, as such, 24 ÷ 4 = 6 (utensils in each group).

31. During our family's 35 days vacation, we will spend 17 days on a ocean cruise. The rest of the time we will divide equally between our 2 grandparents. How many days will we spend with each grandparent? *Ans:* 9 days.
Solution: 35 days - 17 days (on the ocean) = 18 days (with both grandparents). Then, 18 days ÷ 2 = 9 days (with each grandparent).

32. There are 3 tennis balls in each can. Joe took the balls out of 4 cans and then divided them between 3 teams. How many balls did each team get? *Ans:* 4 balls.
    Solution: 3 (tennis balls in each can) × 4 (cans) = 12 (tennis ball in all four cans). Then, 12 (balls) ÷ 3 (teams) = 4 (balls for each team).

33. A coach took her 6 best runners from 4 classes and divided them into 3 teams. How many students were in each team? *Ans:* 8 students.
    Solution: The coach took 6 × 4 = 24 (runners). Then she divided all the runners 24 (runners) ÷ 3 (groups) = 8 (runners in each group).

34. On the table, there are 5 apples in a bowl and also 2 bags with 3 apples in each bag. How many apples are on the table? *Ans:* 11 apples.
    Solution: To solve this problem we need to find first how many apples are in the bags, 3 apples (in each bag) × 2 (bags) = 6 (apples in two bags). Next, 5 apples (in the bowl) + 6 apples (in the bags) = 11 apples (on the table).

35. An army general earned 47 medals. He put 29 medals on his parade uniform and the rest he stored in 3 boxes. How many medals are in each box? *Ans:* 6 medals.

36. On day one, the doctor saw 5 patients. The second day he saw twice as many and on the third day he saw three times as many as the first day. How many patients did the doctor see in three days? *Ans:* 30 patients.

37. Ivan counted 6 maple trees, twice that many pine trees and 8 fruit trees. How many trees did he count? *Ans:* 26 trees.

38. Sue read 2 books the first week, three times that many the next week and then twice that many the following week. How many books did Sue read in three weeks? *Ans:* 2 + 6 + 4 = 12 books.

39. Cesar said that he saw 6 koi and 8 goldfish fish in the pond. He says he saw twice as many frogs as the fish. How many frogs did he see? *Ans:* 28 frogs.

40. Jill put 1 penny on the first square of a chess board. She put twice that many on square 2 and then doubling again on the third square. She put twice as many again on the fourth square as she did on the third square. How many pennies did she put on four squares? *Ans:* 1 + 2 + 4 + 8 = 15 pennies.

MULTIPLICATION AND DIVISION BY 3 OR 4

Multiplication and division facts must be memorized and there is no way around it. Please go over the facts (multiplication and division tables) several times; first few times in order, then randomly. It is common to memorize some facts right away but get stuck on others. With practice all the facts will stay in memory and there will be no errors.

Go for it!

EXERCISE I

5 × 3 = 15	6 × 3 = 18	7 × 3 = 21	9 × 3 = 27
4 × 4 = 16	4 × 6 = 24	8 × 3 = 24	4 × 8 = 32
3 × 4 = 12	3 × 5 = 15	7 × 4 = 28	9 × 4 = 36
4 × 3 = 12	4 × 7 = 28	8 × 4 = 32	4 × 10 = 40
5 × 4 = 20	3 × 6 = 18	3 × 8 = 24	10 × 3 = 30

EXERCISE II

2 × 3 = 6	5 × 2 = 10	2 × 7 = 14	7 × 3 = 21
4 × 3 = 12	5 × 3 = 15	3 × 3 = 9	3 × 6 = 18
3 × 4 = 12	2 × 9 = 18	8 × 2 = 16	4 × 5 = 20
2 × 6 = 12	6 × 3 = 18	4 × 4 = 16	6 × 4 = 24

8 × 3 = 24	9 × 3 = 27	2 × 9 = 18	4 × 8 = 32
4 × 7 = 28	8 × 3 = 24	2 × 10 = 20	4 × 4 = 16
7 × 3 = 21	4 × 8 = 32	4 × 10 = 40	4 × 3 = 12
3 × 9 = 27	3 × 9 = 27	7 × 4 = 28	6 × 2 = 12
7 × 4 = 28	9 × 4 = 36	8 × 3 = 24	

DIVISION BY 3

30 ÷ 3 = 10	21 ÷ 3 = 7	12 ÷ 3 = 4	6 ÷ 3 = 2
27 ÷ 3 = 9	18 ÷ 3 = 6	9 ÷ 3 = 3	3 ÷ 3 = 1
24 ÷ 3 = 8	15 ÷ 3 = 5		

DIVISION BY 4

40 ÷ 4 = 10	28 ÷ 4 = 7	16 ÷ 4 = 4	8 ÷ 4 = 2
36 ÷ 4 = 9	24 ÷ 4 = 6	12 ÷ 4 = 3	4 ÷ 4 = 1
32 ÷ 4 = 8	20 ÷ 4 = 5		

EXERCISE III

30 ÷ 3 = 10	20 ÷ 4 = 5	18 ÷ 3 = 6	36 ÷ 4 = 9
9 ÷ 3 = 3	24 ÷ 3 = 8	24 ÷ 3 = 8	16 ÷ 4 = 4
12 ÷ 3 = 4	24 ÷ 4 = 6	30 ÷ 3 = 10	8 ÷ 4 = 2
12 ÷ 4 = 3	21 ÷ 3 = 7	32 ÷ 4 = 8	8 ÷ 2 = 4
16 ÷ 4 = 4	27 ÷ 3 = 9	28 ÷ 4 = 7	15 ÷ 3 = 5
15 ÷ 3 = 5	28 ÷ 4 = 7	24 ÷ 3 = 8	21 ÷ 3 = 7
18 ÷ 3 = 6	20 ÷ 4 = 5	27 ÷ 3 = 9	16 ÷ 4 = 4

WORD PROBLEMS

1. Ben planted 12 tulips in 4 rows. How many tulips are in each row? *Ans:* 3 tulips.

2. If 5 CDs are in each box,
   a) How many CDs are in 3 boxes? *Ans:* 15 CDs.
   b) How many CDs are in 2 boxes? *Ans:* 10 CDs.
   c) How many CDs are in 4 boxes? *Ans:* 20 CDs.

3. If 4 water bottles fit in each backpack, how many bottles can you fit in 4 backpacks? *Ans:* 16 water bottles.

4. If 5 baby chicks are in 1 nest, how many baby chicks could be in 4 nests? *Ans:* 20 chicks.

5. If 18 players are divided equally into 3 teams, how many players will be in each team? *Ans:* 6 players.

6. If one car has 4 wheels, how many wheels do 6 cars have? *Ans:* 24 wheels.

7. If one turtle has 4 legs, how many legs do 5 turtles have? *Ans:* 20 legs.

8. If a fish has 5 fins, how many fins will be on 4 fish? *Ans:* 20 fins.

9. If 3 red stars have 15 points, how many points does each star have? *Ans:* 5 points.

10. Three gold stars get 18 points, how many points does each star get? *Ans:* 6 points.

11. If 3 stars have 21 points, how many points does each star have? *Ans:* 7 points. (Have you ever seen a 7-point star? I haven't.)

12. If 4 cubes have 24 sides, how many sides does each cube have? *Ans:* 6 sides.

13. If one truck can carry 4 tons, how much can 6 trucks carry? *Ans:* 24 tons.

14. If a room has 4 walls,
    a) How many walls do 4 rooms have? *Ans:* 16 walls.
    b) How many walls do 6 rooms have? *Ans:* 24 walls.
    c) How many walls do 7 rooms have? *Ans:* 28 walls.

15. If I counted 20 mules' legs, how many mules would I see? *Ans:* 5 mules.

16. If I counted 12 cockroaches' legs, can you guess how many cockroaches I saw? Remember that each cockroach has 6 legs. *Ans:* 2 cockroaches.

17. If each motorcycle has 2 riders, how many people are on 9 motorcycles? *Ans:* 18 people.

18. How many wheels do 6 tricycles have? *Ans:* 18 wheels.

19. How many fangs do 6 snakes have, if each snake has 2? *Ans:* 12 fangs.

20. How many legs do 7 dogs have? *Ans:* 28 legs.

21. How many toes do 5 legs have? *Ans:* 25 toes.

22. How many sides do 9 triangles have? *Ans:* 27 sides.

23. If you divide 21 pins equally among 3 pincushions, how may pins will you stick in each cushion? *Ans:* 7 pins.

24. If 4 Martians have 28 eyes, how many eyes does each Martian have? *Ans:* 7 eyes (3 in front, 3 in the back, and 1 on top).

25. How many legs do 3 octopi have? *Ans:* 24 legs.

26. How many legs do 4 octopi have? *Ans:* 32 legs.

27. How many weeks are in 21 days? *Ans:* 3 weeks.

28. How many weeks are in 28 days? *Ans:* 4 weeks.

29. If all the cows have 5 spots each, then how many spots do 4 cows have? *Ans:* 20 spots.

30. One bunch has 8 flowers.
    a) How many flowers are in 3 bunches? *Ans:* 24 flowers.
    b) How many flowers are in 4 bunches? *Ans:* 32 flowers.

31. If a newsletter has 4 pages, how many pages are in 8 newsletters? *Ans:* 32 pages.

32. If each watch has 3 hands (hour, minute, and second), how many hands do 9 watches have? *Ans:* 27 hands.

33. If there are 9 ducks in one pond, how many ducks could be in 4 ponds? *Ans:* 36 ducks.

34. If there are 8 lamp posts per mile, how many lamp posts are there in 4 miles? *Ans:* 32 lamp posts.

35. A puppy learns 4 new tricks each month.
    a) How many tricks will he know in 6 months? *Ans:* 24 tricks.
    b) How many tricks will he know in 7 months? *Ans:* 28 tricks.
    c) How many tricks will he know in 8 months? *Ans:* 32 tricks.

36. For a race, 3 teams brought 6 runners each. How many pairs of runners competed in the race? *Ans:* 9 pairs.
    Solution: 6 runners per team means that 3 teams will have 6 × 3 = 18 runners.
    To make pairs means to group in two, also the same as dividing by 2. So total numbers of pairs we can out of the 18 runners is 18 divided by 2, which makes 9 pairs.

37. There are 3 tennis balls in a can. Joe took all the balls out of 6 cans and then divided them between 9 teams. How many balls did each team get? *Ans:* 2 balls.

38. There are 6 pens in each box. Ms. Grace took out pens from 4 boxes and divided them among 3 teachers. How many pens did each teacher get? *Ans:* 8 pens.

39. There were 4 rows of roses, 4 roses in each row. A gardener replanted them all into 8 rows. How many roses are in each row now? *Ans:* 2 roses.

40. There are 4 bags of goodies,with 5 goodies in each bag. Judy took them out and put 2 goodies each, in some new bags. How many bags did she use? *Ans:* 10 bags.

41. There are 3 buttons on a phone, in 4 rows. The same buttons can be looked at as divided in 3 columns. How many buttons are in each column? *Ans:* 4 buttons.
    You can either use an old phone to show the rows and columns (newer, modern phones may have too many buttons) or draw a phone panel to illustrate, if the child has hard time visualizing it.

42. Karen has 24 toy cars.
    a) First he arranged all his cars in 2 rows. How many cars were in each row? *Ans:* 12 cars.
    b) Then, he arranged his cars in 3 rows. How many cars were in each row? *Ans:* 8 cars.
    c) Then, he arranged his cars in 4 rows. How many cars were in each row? *Ans:* 6 cars.
    d) Can you guess how many cars will be in each row if there are 6 rows? *Ans:* 4 cars.
    e) How about 8 rows? *Ans:* 3 cars.

43. One room has 3 windows. The other has 4 walls with 2 windows on each wall. How many windows are in both rooms? *Ans:* 11 windows.

    Solution: The order of operations is important. First, we need to find how many windows are in the second room, 2 (windows on each wall) × 4 (the number of walls) = 8 windows. Now, in both rooms 3 windows + 8 windows = 11 windows.

44. Ten regular monsters and 3 two-headed monsters live on an island. How many monster heads altogether are on the island? *Ans:* 16 heads.

    Solution: The order of operations is important. First, let's figure out how many heads all two-headed monsters have, 2 heads × 3 (monsters) = 6 heads. Now, we will count all the heads, 10 heads (regular) + 6 heads (two-headed) = 16 heads altogether.

45. There are 5 cars on one lot. On the other lot, there are 3 rows of cars, 4 cars in each row. How many cars are in both lots? *Ans:* 17 cars.

    Solution: The order of operations is important. First, we will find out how many cars are on the second lot, 4 (cars) × 3 (rows) = 12 cars. Then, 5 cars (on the first lot) + 12 cars (on the second lot) = 17 cars (on both lots).

46. On a small farm, there are 3 chickens and 4 pigs. How many legs do these animals have? *Ans:* 22 legs.

    Solution: The order of operations is important. First, we will count the number of chickens' legs, 2 legs × 3 = 6 legs. Then, we will count the pigs' legs 4 legs × 4 = 16 legs, altogether there are 6 legs + 16 legs = 22 legs.

47. An another small farm had 6 ducks and 3 rabbits. How many legs do these animals have? *Ans:* 24 legs.

    Solution: The order of operations is important. First, we will count the number of ducks' legs, 2 legs × 6 = 12 legs. Then, we will count the rabbits' legs 4 legs × 3 = 12 legs. Now, altogether there are 12 legs + 12 legs = 24 legs.

48. On yet another small farm they had 5 ostriches and 5 zebras. How many legs does that make? *Ans:* 30 legs.
    Solution: The order of operations is important. First, we will count the number of ostriches' legs, 2 legs × 5 = 10 legs. Then, we will count the zebras' legs 4 legs × 5 = 20 legs. Altogether there are 10 legs + 20 legs = 30 legs.

49. Yen had 4 earrings. She received 3 more pairs of earrings for her birthday. How many earrings does she have now if each pair has 2 earrings? *Ans:* 10 earrings.
    Solution: The order of operations is important. First, we will find how many earrings Yen received for her birthday, 2 (ear ring) × 3 pairs = 6 (earrings). Now, the number of earrings she has now, 4 earrings (she had before the birthday) + 6 ear rings (for her birthday) = 10 earrings (now).

50. One folder has 3 articles, each 4 pages long; another folder has 4 articles each 5 pages long. How many pages will the editor read from both folders? *Ans:* 32 pages.

25

MULTIPLICATION AND DIVISION BY 5 OR 10

Multiplying by 0 and 10 is simple. When you multiply a number by zero the answer is always zero! And when you multiply zero by a number, it is also zero.

- $1 \times 0 = 0$
- $5 \times 0 = 0$
- $0 \times 7 = 0$
- $0 \times 55 = 0$
- $23 \times 0 = 0$
- $100 \times 0 = 0$
- 1 million $\times 0 = 0$

Easy! Any number times zero is zero.

Never, never divide by zero! Mathematically speaking, it is against the law.

Multiplying by 10 is easy. Simply imagine a zero appearing on the right side of the number.

$1 \times 10 = 10$	$5 \times 10 = 50$	$10 \times 10 = 100$
$2 \times 10 = 20$	$10 \times 7 = 70$	$12 \times 10 = 120$

Dividing by 10 is as easy as removing a zero from the right end of the number. Look!

$10 \div 10 = 1$	$50 \div 10 = 5$
$30 \div 10 = 3$	$100 \div 10 = 10$

EXERCISE I

- Count from 0 to 100 by 10 (i.e., 10, 20, 30, 40).
- Count down from 100 to 10 by 10 (i.e., 100, 90, 80, 70).
- Count from 0 to 55 by 5 (i.e., 0, 5, 10, 15, 20, 25, 30).
- Count down from 95 to 40 by 5 (i.e., 95, 90, 85, 80, 75, 70).

EXERCISE II

$7 + 7 + 7 = 21$	$9 + 9 + 9 = 27$	$15 + 15 + 15 = 45$
$8 + 8 + 8 = 24$	$10 + 10 + 10 = 30$	$16 + 16 + 16 = 48$
$9 + 9 + 9 = 27$	$20 + 20 + 20 = 60$	$17 + 17 + 17 = 51$
$5 + 5 + 5 = 15$	$30 + 30 + 30 = 90$	$25 + 25 + 25 = 75$
$4 + 4 + 4 = 12$	$11 + 11 + 11 = 33$	$30 + 30 + 30 = 90$
$6 + 6 + 6 = 18$	$12 + 12 + 12 = 36$	$24 + 24 + 24 = 72$
$8 + 8 + 8 = 24$	$13 + 13 + 13 = 39$	$25 + 25 + 25 = 75$

EXERCISE III

$1 \times 5 = 5$	$5 \times 5 = 25$	$9 \times 5 = 45$
$2 \times 5 = 10$	$6 \times 5 = 30$	$10 \times 5 = 50$
$3 \times 5 = 15$	$7 \times 5 = 35$	
$4 \times 5 = 20$	$8 \times 5 = 40$	

EXERCISE IV

$8 \div 4 = 2$	$5 \times 4 = 20$	$18 \div 3 = 6$	$27 \div 3 = 9$
$9 \div 3 = 3$	$2 \times 9 = 18$	$6 \times 3 = 18$	$6 \times 5 = 30$
$4 \times 3 = 12$	$3 \times 7 = 21$	$6 \times 4 = 24$	$16 \div 2 = 8$
$5 \times 3 = 15$	$16 \div 4 = 4$	$28 \div 4 = 7$	$7 \times 3 = 21$

$24 \div 4 = 6$	$32 \div 4 = 8$	$24 \div 4 = 6$	$3 \times 6 = 18$
$24 \div 3 = 8$	$3 \times 9 = 27$	$4 \div 4 = 1$	$3 \times 7 = 21$
$4 \times 5 = 20$	$9 \times 4 = 36$	$7 \times 4 = 28$	$16 \div 4 = 4$
$3 \times 5 = 15$	$32 \div 4 = 8$	$3 \times 7 = 21$	

WORD PROBLEMS

1. Howard can jump 10 feet. His pet kangaroo can jump 3 times further. How far can the kangaroo jump? *Ans:* 30 feet.

2. A friendly alien from a far away galaxy has 5 heads, each with 3 ears. How many ears does the alien have? *Ans:* 15 ears.

3. An alien from another galaxy has 3 heads, each with 3 ears. How many ears does this alien have? *Ans:* 9 ears.

4. If the alien's hand has 8 nails, how many nails does he have on 2 of his hands? *Ans:* 16 nails.

5. The alien's foot has 4 toes.
    a) How many toes are on 4 feet? *Ans:* 16 toes.
    b) How many toes are on 5 feet? *Ans:* 20 toes.
    c) How many toes are on 6 feet? *Ans:* 24 toes.
    d) How many toes are on 7 feet? *Ans:* 28 toes.

6. One week has 5 working days.
    a) How many working days are in 3 weeks have? *Ans:* 15 days.
    b) How many working days are in 5 weeks have? *Ans:* 25 days.
    c) How many working days are in 8 weeks have? *Ans:* 40 days.
    d) How many working days are in 9 weeks have? *Ans:* 45 days.

7. If one fish has 5 fins, how many fins do 5 fish have?
    Ans: 25 fins.

8. A watermelon cost $4. How much did Larry pay for 5 watermelons? *Ans:* $20.

9. Twenty boxes of cookies were equally divided among 5 Girl Scouts. How many boxes did each girl get? *Ans:* 4 boxes.

10. If one tractor has 5 wheels, how many wheels do 6 tractors have? *Ans:* 30 wheels.

11. A 30-mile highway was divided among 5 cleaning crews. How many miles of highway did each crew get assigned?
    Ans: 6 miles.

12. There are 5 lamps hanging on each branch. The tree has 8 branches. How many lamps are hanging on the tree?
    Ans: 40 lamps.

13. If there are 40 lemons on a tree and each branch has 5 lemons, how many branches with lemons are there on the tree?
    Ans: 8 branches.

14. The Olympic Games flag has 5 rings.
    a) How many rings do 7 flags have? *Ans:* 35 rings.
    b) How many rings do 9 flags have? *Ans:* 45 rings.
    c) How many rings do 6 flags have? *Ans:* 30 rings.

15. If at a BBQ party each child ate 5 hot dogs, how many hot dogs did 9 children eat? *Ans:* 45 hot dogs.

16. If at the same BBQ party each adult ate 5 hamburgers, how many hamburgers did 10 adults eat? *Ans:* 50 hamburgers.

17. There were 45 exam papers divided equally among 5 teachers. How many exam papers did each teacher get? *Ans:* 9 papers.

18. A decade means 10 years. If a person lives 80 years, how many decades does he live? *Ans:* 8 decades.

19. If 6 fire fighters ride on 1 fire truck, how many can go on 5 trucks? *Ans:* 30 fire fighters.

20. Trisha bought 30 cans of dog food and divided them into 5 bags. How many cans were in each bag? *Ans:* 6 cans.

21. A truck can carry 5 tons of weight. How much load can 5 trucks carry? *Ans:* 25 tons.

22. A tree grew 25 inches in 5 months. How many inches did it grow each month? *Ans:* 5 inches.

23. If there are 20 rabbits altogether inside 4 pens, how many rabbits are in each pen? *Ans:* 5 rabbits.

24. A school band of 35 students went on a field trip in 5 vans. How many students were on each van? *Ans:* 7 students.

25. How many dollars are in six $5 bills? *Ans:* $30.

26. How many dollars are in 10 $5 bills? *Ans:* $50.

27. How many dollars are in 5 $10 bills? *Ans:* $50.

28. A group of 70 students was divided into 10 teams. How many students are in each team? *Ans:* 7 students.

29. If it takes 5 minutes for everyone in my family to take a shower, how long will it take a 6-person family to take their showers? *Ans:* 30 minutes.

30. Tao got 30 stickers to put on 5 packages. How many stickers went on each package? *Ans:* 6 stickers.

31. If it took 45 minutes for Yin to run 5 miles, how long does it take her to run 1 mile? *Ans:* 9 minutes.

32. A boss gave a $10 bonus to 5 of her workers. How much bonus did she give to all? *Ans:* $50.

33. At 8 cents each, how much must be paid for 5 lollypops? *Ans:* 40 cents.

34. In 5 classrooms there are 40 desks. If the rooms have the same number of desks, how many desks are in each room? *Ans:* 8 desks.

35. If a family paid $45 to rent a boat for 5 hours. How much did each hour of boat rental cost? *Ans:* $9.

36. If each book cost $5, how many could be purchased for $25? *Ans:* 5 books.

37. If you bought 5 balloons and paid 50 cents, how much did you pay for each balloon? *Ans:* 10 cents.

38. There were 4 children on the team. Each child ran 5 miles. How many miles did they run all together? *Ans:* 20 miles.

39. Nina has 1 doll. Caitlin has 10 times as many. Caitlin has how many dolls? *Ans:* 10 dolls.

40. If 35 flowers are divided into 5 bunches, how many flowers are in each bunch? *Ans:* 7 flowers.

41. When 5 castaways evenly divided 40 coconuts among themselves, how many did each get? *Ans:* 8 coconuts.

42. When 5 castaways evenly split 45 bananas, how many did each get? *Ans:* 9 bananas.

43. When 5 hungry castaways evenly divided 35 lizards, how many did each get? *Ans:* 7 lizards. What are they going to do with the lizards? Eat them?

44. Mrs. Pie always puts 5 cherries on a sundae.
a) How many cherries does she put on 7 sundaes?
Ans: 35 cherries.
b) How many cherries does she put on 5 sundaes?
Ans: 25 cherries.
c) How many cherries does she put on 6 sundaes?
Ans: 30 cherries.

45. In the village there are 5 dogs for every cat. How many cats are in the village, if it has 45 dogs? *Ans:* 9 cats.

46. If one praying mantis has 6 legs, how many legs do 5 praying mantises have? *Ans:* 30 legs.

47. There are 5 working days in one week.
a) How many working days are in 7 weeks? *Ans:* 35 days.
b) How many working days are in 8 weeks? *Ans:* 40 days.
c) How many working days are in 9 weeks? *Ans:* 45 days.

48. One truck carries 6 tons of goods and 4 loaded trucks delivered goods to a port where it was evenly divided among 3 boats. How much load did each boat carry? *Ans:* 8 tons.
Solution: First find the total number of tons that are there. 6 tons (each truck) × 4 (trucks) = 24 tons (all trucks brought together). Next, 24 tons was divided among 3 boats, 24 tons ÷ 3 = 8 tons (in each boat).

49. Three friends earned $50 for yard work. After paying $23 for the new equipment they divided the rest among themselves. How much did each of them make? *Ans:* $9.
Solution: Together they earned $50 - $23 (for equipment) = $27. Then $27 ÷ 3 = $9 (each).

50. If one car has 5 tires, how many tires do 6 cars have?
Ans: 30 tires.

MULTIPLICATION AND DIVISION BY 6

EXERCISE I

- Count from zero to 60 skipping 3 (0, 3, 6, 9, 12, etc.)
- Count from zero to 60 skipping 6 (0, 6, 12, 18, etc.)

MULTIPLICATION BY 6

$6 \times 1 = 6$	$6 \times 5 = 30$	$6 \times 9 = 54$
$6 \times 2 = 12$	$6 \times 6 = 36$	$6 \times 10 = 60$
$6 \times 3 = 18$	$6 \times 7 = 42$	
$6 \times 4 = 24$	$6 \times 8 = 48$	

DIVISION BY 6

$12 \div 6 = 2$	$30 \div 6 = 5$	$48 \div 6 = 8$
$18 \div 6 = 3$	$36 \div 6 = 6$	$54 \div 6 = 9$
$24 \div 6 = 4$	$42 \div 6 = 7$	$60 \div 6 = 10$

EXERCISE II

Ask child to do the multiplication first, then subtract the second number. Ask: What is six less than six times six?

$6 \times 2 - 6 = 6$	$6 \times 4 - 6 = 18$	$6 \times 6 - 12 = 24$
$6 \times 3 - 6 = 12$	$6 \times 5 - 12 = 18$	$6 \times 7 - 12 = 30$

$6 \times 8 - 24 = 24$ $6 \times 10 - 24 = 36$ $6 \times 3 - 12 = 6$

$6 \times 9 - 24 = 30$ $6 \times 4 - 12 = 12$ $6 \times 6 - 12 = 24$

EXERCISE III

$4 \times 3 = 12$	$3 \times 8 = 24$	$6 \times 6 = 36$	$6 \times 6 = 36$
$2 \times 6 = 12$	$4 \times 6 = 24$	$8 \times 4 = 32$	$9 \times 6 = 54$
$3 \times 6 = 18$	$6 \times 5 = 30$	$4 \times 9 = 36$	$8 \times 6 = 48$
$9 \times 2 = 18$	$4 \times 7 = 28$	$6 \times 7 = 42$	$6 \times 6 = 36$
$6 \times 4 = 24$	$9 \times 3 = 27$	$5 \times 7 = 35$	$6 \times 7 = 42$
$4 \times 4 = 16$	$5 \times 5 = 25$	$6 \times 8 = 48$	$6 \times 5 = 30$

EXERCISE IV

$36 \div 6 = 6$	$24 \div 4 = 6$	$36 \div 6 = 6$	$54 \div 6 = 9$
$12 \div 6 = 2$	$24 \div 6 = 4$	$36 \div 4 = 9$	$60 \div 6 = 10$
$12 \div 4 = 3$	$24 \div 3 = 8$	$42 \div 6 = 7$	$28 \div 4 = 7$
$18 \div 2 = 9$	$30 \div 6 = 5$	$48 \div 6 = 8$	
$18 \div 6 = 3$	$28 \div 4 = 7$	$27 \div 3 = 9$	
$21 \div 3 = 7$	$32 \div 4 = 8$	$32 \div 4 = 8$	

WORD PROBLEMS

1. If there were 6 children on each team,
   a) How many were on 2 teams? *Ans:* 12 children.
   b) How many were on 3 teams? *Ans:* 18 children.
   c) How many were on 5 teams? *Ans:* 30 children.

2. Each pack has 6 bottles of cola.
   a) How many bottles are in 4 packs? *Ans:* 24 bottles.
   b) How many bottles are in 3 packs? *Ans:* 18 bottles.
   c) How many bottles are in 2 packs? *Ans:* 12 bottles.

3. If there are 2 pictures on each page,
   a) How many pictures are on 12 pages? *Ans:* 24 pictures.
   b) How many pictures are on 6 pages? *Ans:* 12 pictures.

4. If an ant has 6 legs, how many legs do 7 ants have?
   Ans: 42 legs.

5. If one aunt has 2 legs, how many legs do 6 aunts have?
   Ans: 12 legs.

6. If one uncle has 2 ankles, how many ankles do 12 uncles have?
   Ans: 24 ankles.

7. When twins got 6 chores each, how many chores did they get
   altogether? *Ans:* 12 chores.

8. If you put 12 mugs in 2 boxes, how many mugs will be in each
   box? *Ans:* 6 mugs.
   a) If you put 12 mugs in 3 boxes, how many will be in each box?
   Ans: 4 mugs.
   b) If you put 12 mugs in 4 boxes, how many will be in each
   box? *Ans:* 3 mugs.
   c) If you put 12 mugs in 6 boxes, how many will be in each box?
   Ans: 2 mugs.

9. If 12 forks are divided equally among 6 guests, how many forks
   would each guest get? *Ans:* 2 forks.

10. How many skateboards and sunglasses do 6 skiers need?
    Ans: 6 skateboards and 6 pairs of sunglasses.

11. How many rackets do 6 tennis players play with?
    Ans: 6 rackets, of course.

12. If 12 chores are divided between 2 kids, how many chores will
    each get? *Ans:* 6 chores.

13. If 12 chores are divided between 6 kids, how many chores will
    each get? *Ans:* 2 chores.

14. If a year, which has 12 months, is divided in four parts called
    quarters, how many months will each part have?
    Ans: 3 months.

15. Paula's father gave her 6 things to do. She told him that she
    plans to finish them all in 3 days. How many things does she
    have to do each day to get everything done in 3 days?
    Ans: 2 things.

16. If a 24-hour day is divided into 4 shifts, how many hours will
    be in each shift? *Ans:* 6 hours.

17. If a 24-hour day is divided into 6 shifts, how many hours will be in each shift? *Ans:* 4 hours.

18. Pam found 3 nests with 6 eggs in each nest. How many eggs did she find? *Ans:* 18 eggs.

19. There are 6 statues in the park with 4 pigeons on each statue. How many pigeons are on all the statues? *Ans:* 24 pigeons.

20. Marina planted 24 tomatoes in 3 rows. How many tomatoes were planted in each row? *Ans:* 8 tomatoes.

21. Sophia divided 24 books evenly between 4 shelves. How many books are on each shelf? *Ans:* 6 books.

22. If Len divided 24 books evenly between 3 shelves, how many books did he put on each shelf? *Ans:* 8 books.

23. If Boris divided 24 books evenly between 6 shelves, how many books did he put on each shelf? *Ans:* 4 books.

24. If a cow gives 6 liters of milk a day, how many liters of milk will it give in 6 days? *Ans:* 36 liters.

25. At the show, 6 clowns juggle 4 balls each. How many balls are being juggled? *Ans:* 24 balls.

26. If there are 4 ponds with 6 alligators in each pond, how many alligators are there? *Ans:* 24 alligators.

27. If Katie sold 28 glasses of lemonade in 4 hours, how many glasses was she selling each hour? *Ans:* 7 glasses.

28. One shift lasts 6 hours.
    a) How long do 4 shifts last? *Ans:* 24 hours or one whole day.
    b) How long do 5 shifts last? *Ans:* 30 hours.
    c) How long do 6 shifts last? *Ans:* 36 hours.

29. One shift lasts 6 hours.
    a) How many shifts are in 30 hours? *Ans:* 5 shifts.
    b) How many shifts are in 24 hours? *Ans:* 4 shifts.
    c) How many shifts are in 36 hours? *Ans:* 6 shifts.

30. In a museum, there are 6 pictures on each wall.
    a) How many pictures are on 2 walls? *Ans:* 12 pictures.
    b) How many pictures are on 4 walls? *Ans:* 24 pictures.
    c) How many pictures are on 8 walls? *Ans:* 48 pictures.

31. There are 6 golf balls in each box.
    a) How many golf balls are in 3 boxes? *Ans:* 18 golf balls.
    b) How many golf balls are in 6 boxes? *Ans:* 36 golf balls.
    c) How many golf balls are in 9 boxes? *Ans:* 54 golf balls.

32. A city has 24 acres of land to be equally divided into lots.
    a) Divided into 6 lots, how big will each lot be? *Ans:* 4 acres.
    b) Divided into 4 lots, how big will each lot be? *Ans:* 6 acres.
    c) If each lot is 3 acres, how many lots will it make? *Ans:* 8 lots.

33. There are 36 sailors to be divided between ships.
    a) How many sailors are on each ship if there are 4 ships?
    Ans: 9 sailors.
    b) How many sailors are on each ship if there are 9 ships?
    Ans: 4 sailors.
    c) How many sailors are on each ship if there are 6 ships?
    Ans: 6 sailors.

34. Rover always stops at a fire hydrant on his walk. Bob and Rover walked on 6 streets with 8 hydrants on each street. At how many hydrants did Rover stop? *Ans:* 48 hydrants.

35. Rover barks at every cat in the neighborhood. With 5 cats on each of the 6 streets, how many times does he barks on his walks? *Ans:* 30 times.

36. A jeweler uses 6 diamonds in one ring, how many diamonds will he need to make 7 rings? *Ans:* 42 diamonds.

37. Six identical necklaces have with 7 black pearls and 5 white pearls each. How many black pearls are in all of them?
    Ans: 42 pearls.

38. Altogether, the painter brought 42 gallons of paint to paint 6 buildings. How many gallons did he use on each building?
    Ans: 7 gallons.

39. The painter needs 3 gallons for each room. He has 19 gallons of paint. If he has to paint 6 rooms, does he have enough paint to finish the job? Will there be any paint left? *Ans:* Yes, he has enough. 1 gallon will be left.

40. A 4-year-old asked 48 questions in 6 minutes. How many questions did she ask per minute? *Ans:* 8 questions. She seems like a very fast talker! I don't think she waits for an answer!

41. It takes 6 painters to paint one mural. How many painters would it take to paint 5 murals? *Ans:* 30 painters. A mural is a large painting usually shown in a public place such as a bank, library or Government building.
Do you know of any murals in your town?

42. If it takes 6 cops to catch a robber, how many cops would it take to catch 7 robbers? *Ans:* 42 cops.

43. If a pair of earphones cost $9, how much would 6 earphones cost? *Ans:* $54.

44. If the carpenter needs 4 posts for each section of the deck and the deck has six sections, how many posts will the carpenter need? *Ans:* 24 posts.

45. Each section of the fence has 9 boards. We need to build a fence that is 5 sections long. How many boards will we need?
Ans: 45 boards.

46. When 54 chili peppers were divided into 6 bags, how many were in each bag? *Ans:* 9 chili peppers.

47. A pet store received 25 parakeets and 11 canaries. They took all the birds and evenly divided them among 6 cages. How many birds went to each cage? *Ans:* 6 birds.
Solution: The store received 25 + 11 = 36 (birds). They divided the birds 36 (birds) ÷ 6 (cages) = 6 (birds in each cage).

48. There are 4 second grade classes at the school. The coach took 3 students from each class and divided them into 2 teams. How many students are on each team?
Ans: 6 students.
Solution: If 3 students come from each class and there are 4 classes, then 3 (students) × 4 = 12 students. Next, 12 students are divided into 2 teams, 12 ÷ 2 = 6 (students in each team).

49. Missy took 6 cushions from each of her 2 sofas and then divided them equally between 3 sofas. How many cushions are on each sofa now? *Ans:* 4 cushions.

Solution: 6 cushions from 2 sofas is 6 × 2 = 12 (cushions). Next, 12 cushions divided between 3 sofas, 12 ÷ 3 = 4 (cushions on each sofa).

50. If there are 6 toys in 2 boxes, how many toys would be in 3 boxes? *Ans:* 9 toys.
Solution: If there are 6 toys in 2 boxes, then in each box there are 6 ÷ 2 = 3 (toys). Next, 3 boxes with 3 toys in each box are 3 × 3 = 9 (toys).

51. If you pay $10 to buy 5 melons, how many melons can you buy for $14? *Ans:* 7 melons.
Solution: If you pay $10 for 5 melons, then you pay $10 ÷ 5 = $2 (for each melon), in other words 1 melon costs $2. Next $14 will buy you 14 ÷ 2 = 7 melons, because $14 must be divided in into $2 chunks for each melon, and there are 7 two-dollar chunks in $14.

52. Tory picked 17 apples from one tree and 33 from another. She took all the apples to bake 5 apple pies. How many apples will go in each pie? *Ans:* 10 apples.
Solution: Tory picked 17 + 33 = 50 apples. Then, he divided apples into 5 groups (pies), 50 ÷ 5 = 10 (apples in each pie).

53. Simon baked 53 turnovers and ate 5 turnovers himself. The rest he divided among 6 members of his team. Did Simon eat more or less turnovers than each member of his team?
Ans: Simon ate less.
Solution: For the team there were 53 (turnovers) - 5 (turnovers that Simon ate) = 48 turnovers. Next, 48 (turnovers) ÷ 6 (members of the team) = 8 (turnovers for each member). Therefore, Simon, who ate only 5 turnovers, had less than his teammates.

54. In the summer camp there were 6 groups of children, 9 kids in each group. When 19 children left the camp, the rest were divided into 5 groups. How many kids are in each new group?
Ans: 7 children.
Solution: Altogether, there were 9 (kids) × 6 (groups) = 54 (kids). Then, 54 - 19 (children who left the camp) = 35 (children left in the camp). Next, 35 (children) ÷ 5 (new groups) = 7 (children in each new group)

55. After 6 children took 4 tennis balls each from the basket, there were 4 balls left. How many tennis balls were in the basket to start with? *Ans:* 28 balls.
    Solution: Children took 4 (balls) × 6 = 24 (balls). Next, 24 (balls) + 4 (balls that were left) = 28 (balls).

56. Mrs. Bush expected 4 guests and baked 6 rolls for each guest. When 6 guests showed up instead, Mrs. Bush baked 6 extra rolls and divided them among the guests. How many rolls did each guest get? *Ans:* 5 rolls.
    Solution: Mrs. Bush first baked 6 (rolls) × 4 = 24 (rolls). When guests came, they had 24 (already baked rolls) + 6 (extra rolls) = 30 (rolls).

MULTIPLICATION AND DIVISION BY 7

EXERCISE I

- Count from zero to 60 skipping 3 (0, 3, 6, 9, 12, etc.)
- Count from zero to 60 skipping 6 (0, 6, 12, 18, etc.)

MULTIPLICATION BY 7

$7 \times 1 = 7$	$7 \times 4 = 28$	$7 \times 7 = 49$	$7 \times 9 = 63$
$7 \times 2 = 14$	$7 \times 5 = 35$	$7 \times 8 = 56$	$7 \times 10 = 70$
$7 \times 3 = 21$	$7 \times 6 = 42$		

DIVISION BY 7

$70 \div 7 = 10$	$49 \div 7 = 7$	$28 \div 7 = 4$	$14 \div 7 = 2$
$63 \div 7 = 9$	$42 \div 7 = 6$	$21 \div 7 = 3$	$7 \div 7 = 1$
$56 \div 7 = 8$	$35 \div 7 = 5$		

EXERCISE II

$7 \times 1 = 7$	$21 + 7 = 28$	$7 \times 5 = 35$	$42 - 21 = 21$
$14 + 7 = 21$	$14 + 14 = 28$	$35 + 7 = 42$	$35 + 14 = 49$
$7 \times 3 = 21$	$21 + 14 = 35$	$28 + 14 = 42$	$7 + 42 = 49$

49 - 21 = 28	70 - 21 = 49	49 ÷ 7 = 7	49 ÷ 7 = 7
49 + 7 = 56	49 - 21 = 28	56 ÷ 7 = 8	8 × 7 = 56
42 + 14 = 56	14 ÷ 7 = 2	28 ÷ 7 = 4	7 × 10 = 70
56 - 21 = 35	35 ÷ 7 = 5	63 ÷ 7 = 9	7 × 9 = 63
56 - 28 = 28	4 × 7 = 28	7 × 8 = 56	6 × 7 = 42
49 + 21 = 70	42 ÷ 7 = 6	3 × 7 = 21	63 ÷ 7 = 9
35 + 21 = 56	7 × 3 = 21	6 × 7 = 42	7 × 7 = 49
70 - 35 = 35	5 × 7 = 35	56 ÷ 7 = 8	6 × 6 = 36
70 - 7 = 63	7 × 7 = 49	9 × 7 = 63	5 × 5 = 25
49 + 14 = 63	7 × 6 = 42	28 ÷ 7 = 4	
56 + 14 = 70	8 × 7 = 56	42 ÷ 7 = 6	

WORD PROBLEMS

1. How many wheels do 7 tricycles have? *Ans:* 21 wheels.

2. How many legs do 7 goats have? *Ans:* 28 legs.

3. How many fingers and thumbs are on 7 hands?
   Ans: 35 fingers and thumbs.

4. Sally has $6, Uma has 7 times as much. How much money does Uma have? *Ans:* $42.

5. One crystal glass costs $7.
   a) How many glasses can you buy for 28 dollars?
   Ans: 4 glasses.
   b) How many glasses can you buy for 42 dollars?
   Ans: 6 glasses.
   c) How many glasses can you buy for 35 dollars?
   Ans: 5 glasses.

6. When 7 postal workers carried 2 bags of mail each, how many bags did all 7 postal workers carry? *Ans:* 14 bags.

7. Two policemen ate 7 doughnuts each, how many doughnuts did they eat in all? *Ans:* 14 doughnuts.

8. Todd put 6 stamps, 7 cents each, on the envelope. What was the total postage? *Ans:* 42 cents.

9. Grandma made 21 pancakes and divided them equally among her 7 grand children. How many pancakes did each of them get? *Ans:* 3 pancakes.

10. If it takes 4 kids to build a robot, how many kids will it take to build 7 robots? *Ans:* 28 kids.

11. If it takes an athlete 7 minutes to run a mile, how long will it take her to run 5 miles? *Ans:* 35 miles. But probably longer.

12. A climber kept 28 carabiners in 7 pockets. How many carabiners were in each pocket? *Ans:* 4 carabiners. A carabiners is locking ring which can be used to attach ropes.

13. A prisoner trying to escape, made a rope from 7 of his shirts, each 6 feet long. What was the length of the rope?
Ans: 42 feet.

14. There are 56 aspen trees in the park planted evenly among 7 paths. How many aspens are on each path? *Ans:* 8 aspens.

15. If a donkey has 7 neck vertebrae (special bones that run from the head down along our back), how many neck vertebrae do 7 donkeys have? *Ans:* 49 vertebrae.

16. How many days are there in 3 weeks? *Ans:* 21 days.

17. It takes 7 minutes for an automatic car wash to wash a car.
a) How long will it take to wash 5 cars? *Ans:* 35 minutes.
b) How long will it take to wash 6 cars? *Ans:* 42 minutes.
c) How long will it take to wash 8 cars? *Ans:* 56 minutes.

18. It takes 7 minutes for an automatic car wash to wash a car.
a) How many cars can it wash in 21 minutes? *Ans:* 3 cars.
b) How many cars can it wash in 28 minutes? *Ans:* 4 cars.
c) How many cars can it wash in 49 minutes? *Ans:* 7 cars.
d) How many cars can it wash in 63 minutes? *Ans:* 9 cars.

19. If one phone number has 7 digits, how many digits do 8 phone numbers have? *Ans:* 56 digits.

20. If Michael works 7 hours a day, how many hours will he work in 5 days? *Ans:* 35 hours.

21. If one ladybug has 7 spots, how many spots do 9 ladybugs have? *Ans:* 63 spots.

22. If each ladybug has 7 spots and Lucy counted 56 spots, how many lady bugs was she looking at? *Ans:* 8 ladybugs.

23. Zane and his 6 friends carried 2 fishing poles each. How many fishing poles did they all carry? *Ans:* 14 fishing poles.

24. For bait, Zane and his 6 friends had 5 worms each. How many worms did they have altogether? *Ans:* 35 worms.

25. Zane and his 6 friends caught 7 fish each. How many fish did they catch altogether? *Ans:* 49 fish.

26. One week is 7 days,
    a) How many days are there in 4 weeks? *Ans:* 28 days.
    b) How many days are there in 7 weeks? *Ans:* 49 days.
    c) How many days are there in 9 weeks? *Ans:* 63 days.

27. One week is 7 days,
    a) How many weeks are there in 35 days? *Ans:* 5 weeks.
    b) How many weeks are there in 42 days? *Ans:* 6 weeks.
    c) How many weeks are there in 56 days? *Ans:* 8 weeks.

28. If one side of the building has 7 columns, how many columns are on all 4 sides? *Ans:* 28 columns.

29. If one dancer wears 7 bracelets on each arm, how many bracelets do 3 dancers wear? *Ans:* 42 bracelets. Solution: There are two ways to solve this problem.
    First Solution: There are altogether 2 (arms for each dancer) × 3 (dancers) = 6 (arms with bracelets). Next, 7 (bracelets on each arm) × 6 (arms) = 42 bracelets.
    Second Solution: Each dancer has 7 (bracelets on each arm) × 2 (arms) = 14 (bracelets). Then, 14 bracelets (1st dancer) + 14 bracelets (2nd dancer) + 14 bracelets (3rd dancer) = 42 bracelets. In the next book we will learn how to multiply double-digit numbers, but until then we'll be adding.

30. Seven brothers have 2 sisters each. How many sisters do they all have? *Ans:* 2 sisters. Did I trick you? If they are brothers, they share the same 2 sisters.

31. If 8 faraway stars have 7 planets each, how many planets do all the stars have? *Ans:* 56 planets.

32. One military helicopter carries 7 soldiers.
    a) How many soldiers can 9 helicopters carry?
    Ans: 63 soldiers.
    b) How many soldiers can 5 helicopters carry?
    Ans: 35 soldiers.
    c) How many helicopters are needed to carry 42 soldiers?
    Ans: 6 helicopters.
    d) How many helicopters are needed to carry 28 soldiers?
    Ans: 4 helicopters.

33. If 1 pitcher can fill 7 glasses, how many glasses can 10 pitchers fill? *Ans:* 70 glasses.

34. If 3 glasses can fill a pitcher, then how many glasses do 3 pitchers hold? *Ans:* 9 glasses.

35. If it takes a digger 7 days to dig 7 holes, how many holes can he dig in 1 day? *Ans:* 1 hole. Of course, 7 holes in 7 days means digging 1 hole each day.

36. A famous poet writes 7 poems a day. How many days will it take him to write 56 poems? *Ans:* 8 days.

37. If a less famous poet writes 3 poems a day, how many poems can he write in 9 days? *Ans:* 27 poems.

38. A famous poet writes 7 poems a day. How many poems will he write on the 8th day? *Ans:* 7 poems.

39. If the Big Dipper has 7 stars, how many stars do 7 astronomers see in the Dipper? *Ans:* 7 stars, of course. I hope you did not fall for this silly trick. The Big Dipper is also known by the name Ursa Major (Latin for Big Bear). It does indeed have 7 stars, though some are double stars.

40. The major scale in music has 7 notes. How many notes are in 5 scales? *Ans:* 35 notes.

41. There are 7 bowls of ice-cream with 6 bees sitting on each bowl. How many bees are eyeing the ice-cream? *Ans:* 42 bees.

42. Suzanne pressed 5 shirts in one hour and there are 30 shirts to be pressed. How long would it take her to press all 30 shirts? *Ans:* 6 hours.

43. On a trip Neil took 7 pictures. Amy took 5 times as many. How many pictures did they take together? **Ans:** 42 pictures.
    Solution: Amy took 7 (pictures) × 5 = 35 pictures. Both together took 7 (pictures) + 35 (pictures) = 42 (pictures).

44. A boat dealer divided 45 boats evenly among 5 warehouses. Then he sold 2 boats from each warehouse. How many boats does the dealer have now? **Ans:** 35 boats.
    Solution: There are two ways to solve this problem. The first way: There are 45 (boats) ÷ 5 = 9 (boats in each warehouse). Next, 9 (boats) - 2 (boats sold) = 7 (boats left in each warehouse). Lastly, 7 (boats) × 5 (warehouses).
    The other way: The dealer has 5 warehouses and he sold 2 boats from each, 2 (boats) × 5 = 10 (boats sold altogether). Then, 45 (boats) - 10 (boats) = 35 (boats remained).

45. After 5 rainy days we had 3 weeks of sunshine. How many of both rainy and sunny days did we have? **Ans:** 26 days.
    Solution: Don't get confused and mix days and weeks. A week has 7 days, and 3 sunny weeks made 7 (days) × 3 = 21 (days). Altogether, there were 5 (days) + 21 (days) = 26 (days)

46. Frankie was supposed to travel for 6 weeks but came back 12 days earlier. How many days did he travel? **Ans:** 30 days.
    Solution: Frankie was supposed to travel 7 (days in a week) × 6 (weeks) = 42 (days). Instead, he came back in 42 (days) - 12 (days) = 30 (days).

47. After 4 weeks of training and 5 days of rest, a baseball team started the new season. How many days did the training and the rest last? **Ans:** 33 days.
    Solution: The training lasted 7 (days in each week) × 4 (weeks) = 28 (days). The training and the rest lasted 28 (days) + 5 (days) = 33 (days).

48. A secret code was made of 4 loud rings followed by 6 quiet knocks. How many knocks and rings 7 secret agents make? **Ans:** 28 loud rings and 42 quiet knocks.

49. On a CD there are 7 songs which are each 2 minutes long and 2 songs which are each 7 minutes long. How long is the CD?
Ans: 28 minutes.
Solution: First, 2 (minutes) × 7 (songs) = 14 minutes; next, 7 (minutes) × 2 (songs) = 14 (minutes). Then, 14 (minutes) + 14 (minutes) = 28 (minutes).

50. If one raccoon has 7 rings on its tail, how many tail rings will 6 raccoons have? **Ans:** 42 rings.

51. A group of 49 space aliens were on 7 flying saucers. How many aliens were in each flying saucer? **Ans:** 7 aliens per saucer.

28

MULTIPLICATION AND DIVISION BY 8

EXERCISE I

- What is one half of 16? *Ans:* 8
- What is one half of 20? *Ans:* 10
- What is one half of 18? *Ans:* 9
- What is one half of 40? *Ans:* 20
- What is one half of 30? *Ans:* 15
- What is one half of 50? *Ans:* 25
- What is one half of 22? *Ans:* 11
- What is one half of 60? *Ans:* 30
- What is one half of 80? *Ans:* 40
- What is one half of 100? *Ans:* 50
- What is one half of 70? *Ans:* 35
- What is one half of 44? *Ans:* 22
- What is one half of 66? *Ans:* 33
- What is one half of 24? *Ans:* 12
- What is one half of 48? *Ans:* 24

MULTIPLICATION BY 8

8 × 1 = 8	8 × 4 = 32	8 × 7 = 56	8 × 9 = 72
8 × 2 = 16	8 × 5 = 40	8 × 8 = 64	8 × 10 = 80
8 × 3 = 24	8 × 6 = 48		

DIVISION BY 8

80 ÷ 8 = 10	56 ÷ 8 = 7	32 ÷ 8 = 4	16 ÷ 8 = 2
72 ÷ 8 = 9	48 ÷ 8 = 6	24 ÷ 8 = 3	8 ÷ 8 = 1
64 ÷ 8 = 8	40 ÷ 8 = 5		

WORD PROBLEMS

1. How many legs do 8 chairs have? *Ans:* 32 legs.

2. How many socks do Snow White and seven dwarfs wear?
    Ans: 16 socks.

3. A golden retriever had 4 puppies every year. How many puppies
    did she have in 8 years? *Ans:* 32 puppies.

4. Not counting the thumbs, how many fingers are on 8 hands?
    Ans: 32 fingers.

5. One trailer has 8 wheels.
    a) How many wheels do 5 trailers have? *Ans:* 40 wheels.
    b) How many wheels do 7 trailers have? *Ans:* 56 wheels.
    c) How many wheels do 8 trailers have? *Ans:* 64 wheels.

6. How many legs do 8 tarantulas have, if each has 8 legs?
    Ans: 64 legs.

7. One trailer has 8 wheels.
    a) How many trailers will have 32 wheels? *Ans:* 4 trailers.
    b) How many trailers will have 48 wheels? *Ans:* 6 trailers.
    c) How many trailers will have 72 wheels? *Ans:* 9 trailers.

8. A pizza has 8 pieces.
    a) How many pieces are in 3 pizzas? *Ans:* 24 pieces.
    b) How many pieces are in 5 pizzas? *Ans:* 40 pieces.
    c) How many pieces are in 6 pizzas? *Ans:* 48 pieces.
    d) How many pizzas will it take to make 16 pieces?
    Ans: 2 pizzas.

e) How many pizzas will it take to make 32 pieces?
Ans: 4 pizzas.
f) How many pizzas will it take to make 56 pieces?
Ans: 7 pizzas.

9. How many legs do 8 lizards have? **Ans:** 32 legs.

10. How many legs do 8 flies have, if each fly has 6 legs?
Ans: 48 legs.

11. How much do 8 movie tickets cost at $7 tickets each?
Ans: $56.

12. If one banjo has 5 strings, how many strings do 8 banjos have?
Ans: 40 strings.

13. A Russian balalaika has 3 strings. How many strings do 8
balalaikas have? **Ans:** 24 strings.

14. If 32 carrots were divided among 8 rabbits, how many carrots
would each rabbit get? **Ans:** 4 carrots.

15. Nazeera poured 64 ounces of lemonade into 8 glasses. How
many ounces went into each glass? **Ans:** 8 ounces.

16. Christopher feeds peanuts to squirrels giving only 8 peanuts to
each.
a) How many squirrels can he feed with 24 peanuts?
Ans: 3 squirrels.
b) How many squirrels can he feed with 48 peanuts?
Ans: 6 squirrels.
c) How many squirrels can he feed with 80 peanuts?
Ans: 10 squirrels.

17. If scientists decide to send 4 rockets to 8 planets, how many
rockets will they need? **Ans:** 32 rockets. Did you know that
until recently astronomers counted 9 planets? In 2006 the
scientists excluded Pluto from the list of planets, so now there
are only eight of them officially.

18. A chef divided 24 potatoes equally among 8 plates. How many
potatoes did she put on each plate? **Ans:** 3 potatoes.

19. Then the chef divided 40 carrots equally among 8 plates. How
many carrots went to each plate? **Ans:** 5 carrots.

20. Fifty six leaves of spinach were divided among 8 plates. How many leaves ended up on each plate? *Ans:* 7 leaves of spinach. Who's coming for dinner? Rabbits?

21. Finally, 16 drum sticks were equally divided among 8 plates. How many drum sticks were on each plate? *Ans:* 2 drum sticks. a) By the way, how many chickens does it take to get 16 wings? *Ans:* 8 chickens.

22. Mason studies vocabulary and plans to memorize 8 new words every day.
a) How many new words will he learn in 6 days?
Ans: 48 words.
b) How many new words will he learn in 8 days?
Ans: 64 words.
c) How many new words will he learn in 9 days?
Ans: 72 words.

23. If Mason learned 56 words studying 8 words a day, how long did it take him to learn them? *Ans:* 7 days or 1 week.

24. On the cruise, Aunt Judy had 4 meals a day. How many meals did she have altogether if the cruise lasted 8 days?
Ans: 32 meals.

25. Eight explorers found a treasure of 40 gold coins and decided to divide them evenly. How many coins did each get?
Ans: 5 coins.

26. A 56-foot tree trunk was cut into 8 equal pieces. How long was each piece? *Ans:* 7 feet.

27. A magician picked a 52-card deck and took out 4 cards. Then he divided the rest into 8 equal stacks. How many cards were in each stack? *Ans:* It should be 6 cards in each stack, but with magicians one never knows.

28. Adi decided to read a 56-page book in 8 days, reading the same number of pages every day. How many pages will she read daily? *Ans:* 7 pages.

29. Then Adi realized that she has only one week left to finish the 56-page book. How many pages a day does she need to read now? *Ans:* 8 pages.

30. After reading 2 pages of her 56 page book, Adi remembered that because of the play rehearsal she has only 6 days to read the book. How fast does she need to read now?
Ans: 9 pages a day.
Solution: After reading 2 pages, there were 56 (pages) - 2 (pages) = 54 (pages left). Adi has to divide 54 pages into 6 days reading, 54 (pages) ÷ 6 = 9 (pages each day).

31. The coach told 7 students to do 8 pullups each. How many pull ups will they all do? *Ans:* 56 pullups.

32. If it takes 8 ants to carry one twig, how many twigs can 72 ants carry? *Ans:* 9 twigs.

33. If a piece of gum cost 8 cents, how many pieces can I buy for 80 cents? *Ans:* 10 pieces of gum.

34. There was $100 in Nomi's account. She took out some money and then had $68 left. With the money she took out, Nomi bought 8 notebooks. How much did she pay for each notebook?
Ans: $4.
Solution: Nomi took out from the bank $100 - $68 = $32. Then, with $32 she bought 8 notebooks at $32 ÷ 8 = $4 (per notebook).

35. A restaurant received 80 pounds of potatoes and used all of them in 8 days. How many pounds of potatoes did they use each day? *Ans:* 10 pounds.

36. Janice invited 7 friends over and took out 8 plates.
a) Dividing 56 cherries, how many did she put on each plate?
Ans: 7 cherries.
b) Dividing 48 strawberries, how many did she put on each plate? *Ans:* 6 strawberries.
c) Dividing 32 pears, how many did she put on each plate?
Ans: 4 pears.
d) Dividing 24 hot dogs, how many did she put on each plate?
Ans: 3 hot dogs.

37. There were 5 white rabbits in the forest and 8 times as many gray rabbits. How many rabbits of both colors lived there?
Ans: 45 rabbits (5 × 8 = 40; 5 + 40 = 45).

38. There are 4 rows of melons with 8 melons in each row. How many melons are there? *Ans:* 32 melons.

39. There are 8 rows of chairs in the room with 7 chairs in each row. How many chairs are in the room? *Ans:* 56 chairs.

40. There are 8 rows of squares on a chess board with 8 squares in each row. How many squares are on a chess board? *Ans:* 64 squares.

41. There are 8 rows of chairs in the room with 5 chairs in each row. How many chairs are in the room? *Ans:* 40 chairs.

42. There are 6 rows of chairs in the room with 7 chairs in each row. How many chairs are in the room? *Ans:* 42 chairs.

43. A chess board has 8 rows and 8 columns of squares. How many squares does it have? *Ans:* 64 squares.

29

MULTIPLICATION AND DIVISION BY 9

TRICK WITH A RING

You might already know the ring trick for times table with 9. It is quite simple. You can multiply any single-digit number by 9 with your hands and a ring. Holding both hands and counting from the left, mark the finger with a ring. You can either take a key ring or make a ring from paper, ribbon, or a piece of plastic. Then, for the answer, the fingers on the left side of the ring will be in the tens place and those on the right in ones place.

For example, if you want to multiply 9 by 3, put the ring on the third finger. Then, 2 fingers on the left of the ring are the number of tens, and 7 fingers on the right are ones. Thus, $9 \times 3 = 27$.

The other trick for multiplying any number by 9 is to multiply it by 10 and then take away the number from the product.

For example:

$5 \times 9 = ?$
   Step 1: $5 \times 10 = 50$
   Step 2: $50 - 5 = 45$.

$8 \times 9 = ?$

Step 1: $8 \times 10 = 80$

Step 2: $80 - 8 = 72$.

MULTIPLICATION BY 9

$9 \times 1 = 9$	$9 \times 4 = 36$	$9 \times 7 = 63$	$9 \times 9 = 81$
$9 \times 2 = 18$	$9 \times 5 = 45$	$9 \times 8 = 72$	$9 \times 10 = 90$
$9 \times 3 = 27$	$9 \times 6 = 54$		

DIVISION BY 9

$90 \div 9 = 10$	$63 \div 9 = 7$	$36 \div 9 = 4$	$18 \div 9 = 2$
$81 \div 9 = 9$	$54 \div 9 = 6$	$27 \div 9 = 3$	$9 \div 9 = 1$
$72 \div 9 = 8$	$45 \div 9 = 5$		

EXERCISE I

- To which number do you add 9 to make 18? *Ans:* 9
- To which number do you add 9 to make 81? *Ans:* 72
- To which number do you add 9 to make 54? *Ans:* 45
- To which number do you add 9 to make 45? *Ans:* 36
- To which number do you add 9 to make 72? *Ans:* 63
- To which number do you add 9 to make 27? *Ans:* 18
- To which number do you add 9 to make 63? *Ans:* 54
- To which number do you add 18 to make 54? *Ans:* 36
- To which number do you add 36 to make 72? *Ans:* 36
- To which number do you add 27 to make 54? *Ans:* 27
- To which number do you add 45 to make 90? *Ans:* 45
- To which number do you add 27 to make 63? *Ans:* 36
- To which number do you add 63 to make 90? *Ans:* 27

WORD PROBLEMS

1. If one man takes 9 drives,
   a) How many drives do 3 men take? *Ans:* 27 drives.
   b) How many drives do 5 men take? *Ans:* 45 drives.
   c) How many drives do 7 men take? *Ans:* 63 drives.

2. How many wheels do 9 cars have? *Ans:* 36 wheels.

3. How many tusks do 9 elephants have? *Ans:* 18 tusks.

4. If one dinner fork is 9 inches long, how many forks can we line up in 63 inches? *Ans:* 7 forks.

5. How many 9-inch pieces can you cut out of 45 inches of string? *Ans:* 5 pieces.

6. How many 9-inch pieces can one cut out of 72 inches of wire? *Ans:* 8 pieces.

7. How many sleeves do 9 shirts have? *Ans:* 18 sleeves.

8. If there 63 seats equally divided between 9 rows, how many seats are in each row? *Ans:* 7 seats.

9. How many blank sheets of paper did Orlando use to make 6 copies of a 9-page brochure? *Ans:* 54 sheets.

10. When 81 pounds of supplies were divided among 9 hikers, how many pounds did each person get? *Ans:* 9 pounds.

11. The alarm clock has to ring 9 times before Phoebe gets out of bed. How many times did the alarm ring in 7 days? *Ans:* 63 times.

12. A 54-pound load was divided among 9 carts. How many pounds are on each cart? *Ans:* 6 pounds.

13. A band recorded 72 songs on 9 CDs. How many songs were on each CD? *Ans:* 8 songs.

14. If each picture frame costs $9, how many can you buy for $54? *Ans:* 6 frames.

15. Six friends decided to write a book. Each of them wrote a 9-page story and added them together to make a book. How many pages are in the book? *Ans:* 54 pages.

16. In a 7-story building, each floor has 9 apartments. How many apartments are in the building? *Ans:* 63 apartments.

17. A hotel has 9 floors and 72 rooms. How many rooms are on each floor, if every floor has the same number of rooms? *Ans:* 8 rooms.

18. Benny is 9 years old.
    a) Benny's mother is 5 times older. How old is she?
    Ans: 45 years old.
    b) Benny's aunt is 6 times older. How old is she?
    Ans: 54 years old.
    c) Benny's grandma is 7 times older. How old is she?
    Ans: 63 years old.
    d) Benny's grandpa is 8 times older. How old is he?
    Ans: 72 years old.

19. For a ballroom dance competition, 9 schools sent 4 pairs each. How many dancers participated in the competition?
    Ans: 72 dancers.
    Solution 1: There are two ways to solve this problem. We can first find how many dancers each school sent, 2 (dancers - one pair) × 4 (pairs from each school) = 8 (dancers from each school). Next, 8 (dancers from each school) × 9 (schools) = 72 (dancers).
    Solution 2: 4 (pairs from each school) × 9 (schools) = 36 (pairs of dancers). Then, 2 (dancers per each pair) × 36 (pairs) = 72 (dancers).

20. There are 9 tents with 4 campers in each tent at a campground. How many campers are there? *Ans:* 36 campers.

21. That evening 9 shy children and 4 times as many brave, outgoing children came to the school dance. How many brave, outgoing children came? *Ans:* 36 children.

22. The triplets have 3 party dresses each. How many party dresses do all three have? *Ans:* 9 dresses.

23. If one dragon has 9 heads,
    a) How many heads do 3 9-headed dragons have?
    Ans: 27 heads.
    b) How many heads do 5 9-headed dragons have?
    Ans: 45 heads.
    c) How many heads do 7 9-headed dragons have?
    Ans: 63 heads.

24. If you can put only 9 ice cubes in each glass,
    a) How many ice cubes can you fit in 2 glasses?
    Ans: 18 ice cubes.
    b) How many ice cubes can you fit in 4 glasses?
    Ans: 36 ice cubes.
    c) How many ice cubes can you fit in 6 glasses?
    Ans: 54 ice cubes.

25. If you can put only 9 ice cubes in each glass,
    a) How many glasses will you need for 36 ice cubes?
    Ans: 4 glasses.
    b) How many glasses will you need for 72 ice cubes?
    Ans: 8 glasses.
    c) How many glasses will you need for 63 ice cubes?
    Ans: 7 glasses.
    d) How many glasses will you need for 81 ice cubes?
    Ans: 9 glasses.

26. Old palace had 9 columns on each side. How many columns were on all sides? *Ans:* 36 columns.

27. If the building has 20 steps in each staircase and it had 3 staircases, how many total steps does the building have?
    Ans: 60 steps.

28. Defending the fortress, 9 warriors shot 63 arrows. How many arrows did each of them shoot? *Ans:* 7 arrows.

29. A 9-room country inn ordered 18 sheets. How many sheets did they order for each room? *Ans:* 2 sheets.

30. The 9-room country inn also ordered 36 pillow cases. How many pillow cases did they order for each room?
    Ans: 4 pillow cases.

31. The same 9-room inn ordered 54 towels. How many towels did they order for each room? *Ans:* 6 towels.

32. A military uniform has 9 buttons. How many buttons are on 7 uniforms? *Ans:* 63 buttons.

33. If a military uniform has 9 buttons, how many uniforms use 72 buttons? *Ans:* 8 uniforms.

34. If there are 9 candles in a one candle holder, how many candles will be in 9 candle holders? *Ans:* 81 candles.

35. One backpack weighs 9 pounds.
    a) How much do 7 backpacks weigh? *Ans:* 63 pounds.
    b) How much do 9 backpacks weigh? *Ans:* 81 pounds.
    c) How much do 8 backpacks weigh? *Ans:* 72 pounds.

36. If one astronaut has to do 9 test flights, then how many test flights would be taken by 9 astronauts? *Ans:* 81 flights.

37. A beekeeper divided 63 ounces of honey into 9 jars. How many ounces of honey are in each jar? *Ans:* 7 ounces.

38. The maids in a hotel had 54 beds to make. If there were 9 maids, how many beds does each maid make per day?
    Ans: 6 beds.

39. A photographer took pictures of 42 smiling children. If each picture had 6 children, how many photographs did he take?
    Ans: 7 photographs.

40. Roger the Robot has 8 buttons. If Marty pushed each of Roger's buttons 6 times, how many times did Marty push buttons in all? *Ans:* 48 times.

41. If Mr. Ford, the car salesman, sold 9 cars this week, how many cars could he sell in 9 weeks? *Ans:* 81 cars.

42. If another car salesman, Mr. Honda, sold 14 cars in one week, how many cars did he sell each day? *Ans:* 2 cars a day.

43. Kerry counted 63 spots on all the ladybugs in her ladybug collection. If she has only 9 ladybugs, how many spots does each lady bug have? *Ans:* 7 spots.

44. The house has six outside walls. It has two windows on each except the font wall which 4 windows. How many total windows does this house have? *Ans:* 14 windows.

45. I read 9 pages each day. The book I am reading is 81 pages long. How many days will it take me to read this book? *Ans:* 9 days.

46. 9 pairs of boys and girls are in the square dancing competition. How many people are competing in all? *Ans:* 18 people. Remember, a pair is the same as 2 people.

THE END

YOU ARE NOW READY FOR

OUR PRODUCTS
THE READING LESSON

- A step-by-step reading program for children ages 3-7

THE VERBAL MATH LESSON BOOK SERIES

- Verbal Math Lesson Level 1
  Suitable for children in K-1.

- Verbal Math Lesson Level 2
  Suitable for children in grades 1 to 2.

- Verbal Math Lesson Level 3
  Suitable for children in grades 3 to 5.

- Verbal Math Lesson Fractions
  Suitable for children in grades 5 to 7.

- Verbal Metrics - Learning Metric Conversions

- Verbal Math Lesson Percents
  Suitable for children in grades 5 to 7.

For description and prices, please see our websites:
www.readinglesson.com
www.mathlesson.com